AQA GCSE
BIOLOGY
HIGHER
PRACTICE TEST PAPERS

FRAN WALSH

Contents

ACKNOWLEDGEMENTS

The author and publisher are grateful to the copyright holders for permission to use quoted materials and images.

All images are © Shutterstock and © HarperCollins Publishers.

Every effort has been made to trace copyright holders and obtain their permission for the use of copyright material. The author and publisher will gladly receive information enabling them to rectify any error or omission in subsequent editions. All facts are correct at time of going to press.

Published by Letts Educational
An imprint of HarperCollins*Publishers*
1 London Bridge Street
London SE1 9GF

ISBN: 9780008276164

First published 2018

10 9 8 7 6 5 4 3 2 1

© HarperCollins*Publishers* Limited 2018

British Library Cataloguing in Publication Data.

A CIP record of this book is available from the British Library.

Commissioning Editor: Clare Souza
Author: Fran Walsh
Editor and Project Manager: Tracey Cowell
Project Editor: Charlotte Christensen
Cover Design: Amparo Barrera
Inside Concept Design: Ian Wrigley
Text Design and Layout: Contentra Technologies, India
Production: Natalia Rebow
Printed and bound by CPI Group (UK) Ltd, Croydon, CR0 4YY

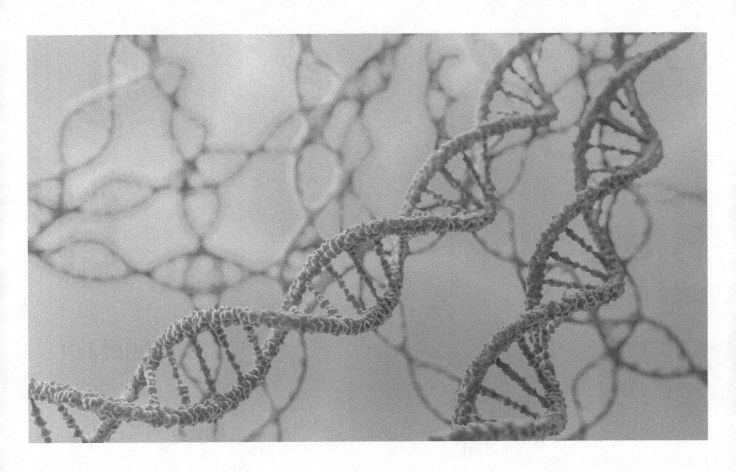

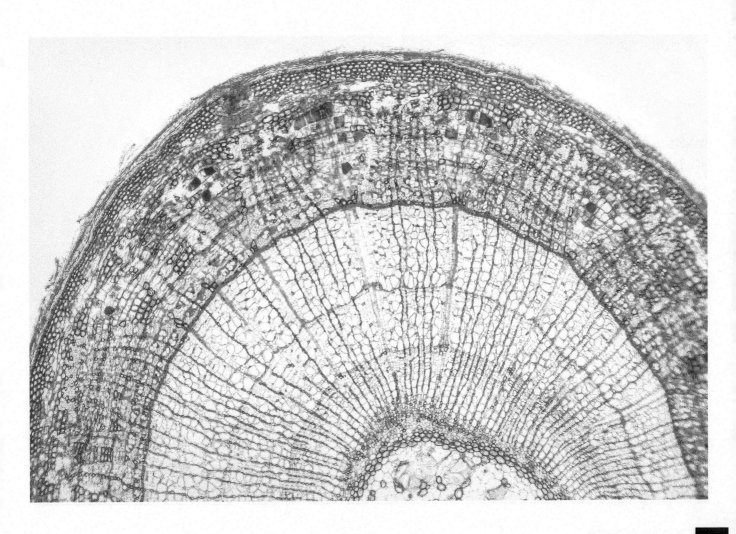

GCSE (9–1)
Biology
Paper 1H

Higher tier

Time: 1 hour 45 minutes

Materials

For this paper you must have:
- a ruler
- a calculator.

Instructions

- Answer **all** questions in the spaces provided.
- Do all rough work in this book. Cross through any work you do not want to be marked.

Information

- There are 100 marks available on this paper.
- The marks for questions are shown in brackets.
- You are expected to use a calculator where appropriate.
- You are reminded of the need for good English and clear presentation in your answers.
- When answering questions 02.1, 05.4, 06.1, 07.1 and 09.4 you need to make sure that your answer:
  - is clear, logical, sensibly structured
  - fully meets the requirements of the question
  - shows that each separate point or step supports the overall answer.

Advice

In all calculations, show clearly how you work out your answer.

Name: _____

0 1 Measles is a disease caused by a virus.

Children are usually given a vaccination to protect them against measles.

0 1 • 1 Describe how the body responds when a vaccinated person encounters the measles virus.

[4 marks]

_____

_____

_____

_____

In 2016, there were 531 cases of measles in the UK. 420 of these cases were in people who had not been vaccinated. A number of these cases were linked to music festivals and other large public events.

0 1 • 2 Calculate the percentage of people who caught measles in 2016 who had not been vaccinated. Give your answer to 2 significant figures. **[2 marks]**

_____

_____

Percentage = _____

0 1 • 3 How is measles spread? **[1 mark]**

Tick **one** box.

By air ☐

By direct contact ☐

By water ☐

By insect vector ☐

0 1 • 4 AIDS is a disease that is also caused by a virus.

State **one** way that the virus causing AIDS is spread. **[1 mark]**

_____

0 1 • 5 Explain why it is difficult to treat diseases such as measles and AIDS with drugs. **[3 marks]**

_____

_____

_____

Figure 1 shows a plant.

Figure 1

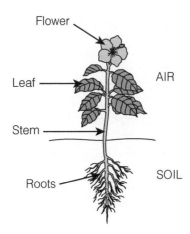

Describe how water moves through the plant from the soil to the air.

Include the names of the processes involved and explain how the different structures of the plant are suited to their function. **[6 marks]**

_____

_____

_____

_____

_____

_____

_____

Some students wanted to investigate water loss in plant leaves.

They took 20 leaves from a plant and weighed them.

They spread the leaves out on a bench and weighed them again every 30 minutes.

Graph 1 shows their results.

Graph 1

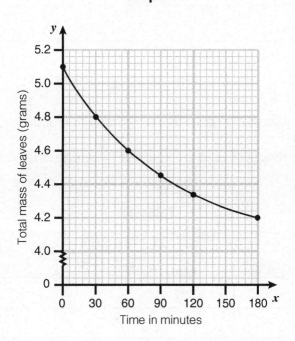

Total mass of leaves (grams) vs Time in minutes

| 0 2 • 2 | Calculate the total mass of water lost by the leaves over the 180 minutes. | **[1 mark]** |

_____

| 0 2 • 3 | Calculate the rate of water loss of the leaves in mg/minute. | **[1 mark]** |

_____

| 0 2 • 4 | Name the process that plants use to take in mineral ions from the soil. | **[1 mark]** |

_____

| 0 2 • 5 | Magnesium is a mineral ion. Plants deficient in this ion suffer from chlorosis. Leaves produce insufficient chlorophyll. | |

Suggest how this will affect the plant. **[1 mark]**

_____

| 0 | 3 | | Enzymes are biological catalysts. |

| 0 | 3 | • | 1 | Enzyme A is added to substrate X and a reaction occurs. |

Enzyme A is added to substrate Y. No reaction occurs.

Suggest why no reaction occurs with substrate Y. **[2 marks]**

_____

_____

_____

Amylase is an enzyme that digests starch to produce glucose.

Table 1 shows the results of an experiment to investigate the effect of pH on the action of amylase.

Table 1

pH	Time taken for starch to be completely digested (minutes)
5	40
6	27
7	16
8	17
9	29
10	42

| 0 | 3 | • | 2 | Plot these results on the graph paper. **[4 marks]**

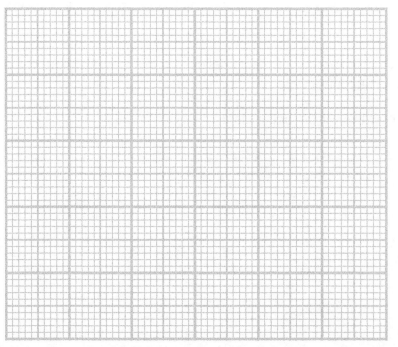

0 3 • 3 What can you conclude from these results? **[1 mark]**

_____

0 3 • 4 Suggest how long it would take for the starch to be completely digested at pH 14. **[1 mark]**

_____

0 3 • 5 Which of the graphs show the results you would expect when investigating the effect of temperature on amylase action? **[1 mark]**

Tick **one** box.

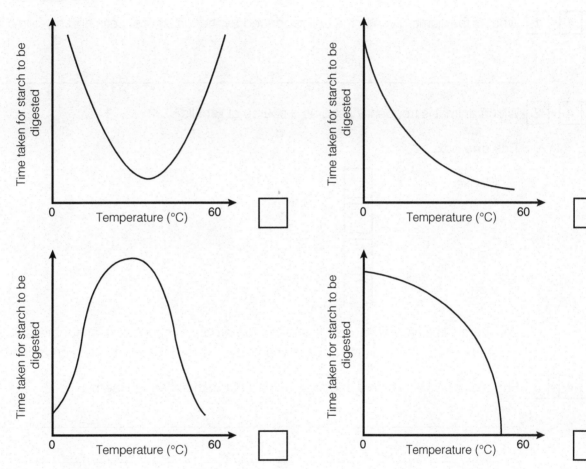

0 4 Read the following facts about monoclonal antibodies.

Special white blood cells called B cells produce antibodies.

B cells cannot be grown outside the body.

Myeloma cells (cancer cells) divide rapidly.

By fusing B cells and myeloma cells, scientists can create a cell that grows in the laboratory, divides rapidly and produces antibodies.

The antibodies produced are called monoclonal antibodies.

0 4 • 1 What is the name given to the cell produced by the fusion of B cells and myeloma cells?

[1 mark]

_____

0 4 • 2 Which animal is frequently used as a source of B cells? **[1 mark]**

Tick **one** box.

Monkey ☐

Mouse ☐

Rabbit ☐

Dog ☐

Leukaemia is a type of cancer. Scientists can produce monoclonal antibodies that will bind to leukaemia cells. They can attach toxic chemicals that destroy cells to the monoclonal antibodies.

0 4 • 3 Suggest why this is a very effective method of treatment for leukaemia. **[2 marks]**

_____

_____

Monoclonal antibodies are used in pregnancy tests. The following passage describes how a pregnancy test works.

The urine moves along absorbent fibres on the pregnancy test stick to zone 1.

In zone 1 there are monoclonal antibodies for HCG (the hormone found in the urine of pregnant women).

The urine carries the antibodies to zone 2 where a chemical reacts with the monoclonal antibodies only if they have HCG attached to them. The reaction causes a colour change from colourless to blue.

The urine now moves to zone 3. Here, another chemical reacts with the monoclonal antibodies even if they do not have HCG bound to them. The reaction causes a colour change from colourless to blue.

| 0 4 • 4 | Complete **Table 2** to show the results you would expect for each scenario. | **[2 marks]** |

Table 2

Scenario	Colour of zone 2	Colour of zone 3
Urine from pregnant woman		
Urine from non-pregnant woman		

| 0 4 • 5 | Explain the purpose of zone 3. | **[1 mark]** |

_____

| 0 5 | This question is about food and digestion. |

| 0 5 • 1 | The food we eat contains proteins, carbohydrates and fats.

Biuret reagent is blue but changes colour in the presence of protein.

What colour would you expect to see if the food contained protein? | **[1 mark]** |

_____

The stomach produces acid that aids digestion. However, the pH in the small intestine is neutral.

| 0 5 • 2 | Explain why the pH in the small intestine is neutral. | **[2 marks]** |

_____

_____

| 0 5 • 3 | Describe the roles of the enzymes lipase, amylase and protease, which are found in the small intestine. | **[3 marks]** |

_____

_____

_____

_____

_____

0 5 • 4 Explain how the small intestine is adapted for absorption of the products of digestion.

[6 marks]

_____

_____

_____

_____

_____

_____

_____

0 6 Respiration occurs in living cells and may be aerobic or anaerobic.

0 6 • 1 Compare aerobic and anaerobic respiration in humans. **[6 marks]**

_____

_____

_____

_____

_____

_____

_____

0 6 • 2 The energy transferred by respiration is used to synthesise new molecules.

Which **two** molecules are needed to synthesise lipids? **[1 mark]**

Tick **one** box.

Glucose and fatty acids ☐

Glycerol and fatty acids ☐

Amino acids and fatty acids ☐

Sucrose and fatty acids ☐

0 6 • 3 Plants use glucose to synthesise the molecule cellulose.

Why do plants need cellulose? **[1 mark]**

_____

0 6 • 4 When synthesised molecules are no longer needed, they are broken down into smaller molecules.

Describe what happens in humans to proteins that are no longer needed.　　　**[3 marks]**

_____

_____

_____

0 6 • 5 Suggest **two** processes in humans, other than synthesis or breakdown of molecules, that use the energy transferred by respiration.　　　**[2 marks]**

_____

0 6 • 6 Yeast is an organism that is able to respire anaerobically.

What name is given to this process?　　　**[1 mark]**

_____

Figure 2 shows the equipment used to make wine.

Fruit, glucose and yeast are placed in the demi-john. Water is placed in the air lock. The apparatus is left for several weeks.

Figure 2

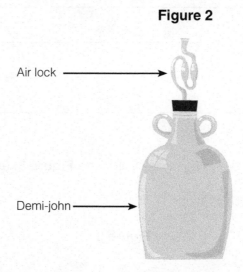

Air lock

Demi-john

0 6 • 7 Complete the equation below to show the process happening in the demi-john.　　　**[1 mark]**

Glucose ──────────▶ _____ + _____

| 0 | 6 | • | 8 | If the temperature is maintained at around 20°C, the wine will have a rich fruity flavour but will take a long time to produce. Raising the temperature to 35°C will produce wine faster but with less taste.

Suggest what will happen if the temperature is raised above 50°C. Explain your answer.

[2 marks]

_____

_____

| 0 | 7 | This question is about cells.

| 0 | 7 | • | 1 | Describe the differences in the structure of prokaryotes and eukaryotes, and give an example of each.

[6 marks]

_____

_____

_____

_____

_____

_____

_____

_____

_____

Some students were observing cheek cells (see **Figure 3**) using a light microscope.

They recorded their observations.

Figure 3

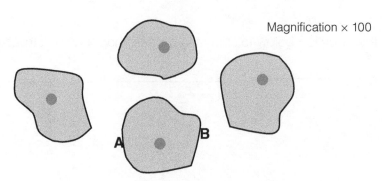

Magnification × 100

Using a micrometer, they measured the width of a cell between points A and B as 5.8 mm.

0 7 . 2 Calculate the actual width of the cell. Give your answer in μm. **[1 mark]**

_____

0 7 . 3 What type of cells are cheek cells? **[1 mark]**

Tick **one** box.

Epithelial cells ☐

Muscle cells ☐

Ciliated cells ☐

Goblet cells ☐

0 7 . 4 Cheek cells divide in a series of stages called the cell cycle.

In the first stage of the cell cycle, the cell grows and increases the number of sub-cellular structures.

Explain what is happening in **Figure 4**. **[1 mark]**

Figure 4

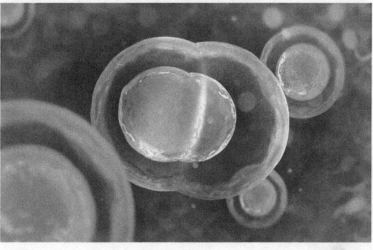

_____

0 7 . 5 Explain why this type of cell division is important in animals. **[1 mark]**

_____

Cells become specialised to carry out particular functions.

0 7 . 6 Explain how a muscle cell is adapted to its function. **[2 marks]**

_____

_____

| 0 | 8 | | | Substances can move in and out of cells by diffusion.

| 0 | 8 | • | 1 | What is diffusion?   **[2 marks]**

_____

_____

The structure shown in **Figure 5** is found in the lungs.

Figure 5

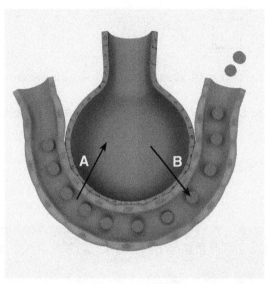

| 0 | 8 | • | 2 | Name the substances represented by arrows A and B.   **[1 mark]**

A _____   B _____

| 0 | 8 | • | 3 | Explain how the structure of the lungs / alveoli are adapted to maximise the rate of diffusion.
                                                                      [4 marks]

_____

_____

_____

_____

Some students wanted to investigate how temperature affected diffusion.

They set up the following apparatus (**Figure 6**) and placed it in a water bath at 10°C.

Figure 6

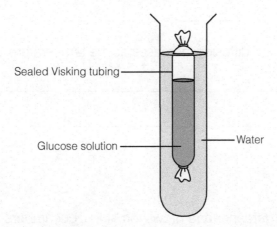

After every minute, for eight minutes, they dipped a glucose testing stick into the water.

The glucose testing stick changes from blue to brown when a certain concentration of glucose is reached.

The students repeated the test at 15°C, 20°C and 25°C.

| 0 | 8 | • | 4 | Suggest **two** factors the students needed to control to make the test fair. **[2 marks]**

_____

_____

Table 3 shows the students' results.

Table 3

Time (minutes)	Colour of glucose testing stick at different temperatures			
	10°C	**15°C**	**20°C**	**25°C**
0	Blue	Blue	Blue	Blue
1	Blue	Blue	Blue	Blue
2	Blue	Blue	Blue	Brown
3	Blue	Blue	Brown	Brown
4	Blue	Blue	Brown	Brown
5	Blue	Brown	Brown	Brown
6	Brown	Brown	Brown	Brown
7	Brown	Brown	Brown	Brown
8	Brown	Brown	Brown	Brown

0 8 • 5 One student concluded that doubling the temperature doubles the rate of diffusion. What data from the table supports this conclusion? **[1 mark]**

_____

_____

0 8 • 6 Explain why the rate of diffusion increases as the temperature increases. **[1 mark]**

_____

_____

0 9 Jeremiah is a keen gardener and grows tomatoes, cucumbers and peppers, which he sells at a local market.

Over the past couple of days he has been worried that some of his tomato plants are diseased.

0 9 • 1 State two symptoms that Jeremiah might have noticed that caused him to suspect a disease. **[2 marks]**

_____

_____

Jeremiah asks a neighbouring farmer to look at his plants. The farmer thinks they may be suffering from infection with tobacco mosaic virus (TMV).

0 9 • 2 State one way that Jeremiah could confirm this diagnosis. **[1 mark]**

_____

Jeremiah asks a friend to look at his plants. She thinks they are nitrate deficient.

0 9 • 3 Explain why plants need nitrates. **[1 mark]**

_____

TMV enters plants through wounds after human handling or animal damage.

0 9 • 4 Describe and explain the variety of mechanisms that plants use to defend themselves against disease. **[6 marks]**

_____

_____

_____

_____

_____

_____

0 9 • 5 Look at **Figure 7**, which shows leaves that have come from a plant with rose black spot disease.

Figure 7

What type of organism is the cause of rose black spot disease? **[1 mark]**

_____

END OF QUESTIONS

GCSE (9–1)
Biology
Paper 2H

Higher tier

Time: 1 hour 45 minutes

Materials

For this paper you must have:
- a ruler
- a calculator.

Instructions

- Answer **all** questions in the spaces provided.
- Do all rough work in this book. Cross through any work you do not want to be marked.

Information

- There are 100 marks available on this paper.
- The marks for questions are shown in brackets.
- You are expected to use a calculator where appropriate.
- You are reminded of the need for good English and clear presentation in your answers.
- When answering questions 02.5, 04.3, 08.2 and 10.4 you need to make sure that your answer:
  - is clear, logical, sensibly structured
  - fully meets the requirements of the question
  - shows that each separate point or step supports the overall answer.

Advice

In all calculations, show clearly how you work out your answer.

Name: _____

0 1 Look at **Figure 1** of a deep-sea anglerfish.

Figure 1

Anglerfish can be found deep in the ocean where light does not penetrate and it is permanently dark. There are no photosynthetic plants so the fish prey on a wide range of other creatures that live in the same environment.

The anglerfish is usually grey in colour and has a rounded body. The shape is designed for remaining motionless rather than swimming. Like most fish, its eyes are on the side of its head.

Classification of the anglerfish places it in the order *Lophiiformes*. *Lophiiformes* are bony fish, characterised by having a 'glowing bulb' that hangs from the top of the head.

0 1 • 1 Suggest how the glowing bulb helps the anglerfish to survive in its environment.   **[2 marks]**

_____

In 2009, a frog fish was discovered in the Indian Ocean. This fish is brightly coloured and appears to hop along the seabed using its fins. It is unusual because it has forward-facing eyes. It does not have a 'glowing bulb' like the anglerfish.

Scientists have classified the new fish as belonging to the order *Lophiiformes*.

0 1 • 2 What evidence may scientists have used to arrive at this classification?   **[1 mark]**

_____

The new fish has been named *Histiophryne psychedelica* using the binomial system.

0 1 • 3 Which part of the name represents the genus?   **[1 mark]**

_____

0 1 • 4 Every year, many new species are discovered.

What is meant by the term species?   **[1 mark]**

_____

0 1 . 5 The classification system developed by Linnaeus has a number of levels.

Which level contains the largest number of organisms? **[1 mark]**

Tick **one** box.

Family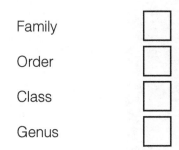

Order

Class

Genus

Evolutionary trees can be used to show the relationships between organisms.

The evolutionary tree in **Figure 2** shows some vertebrates.

Figure 2

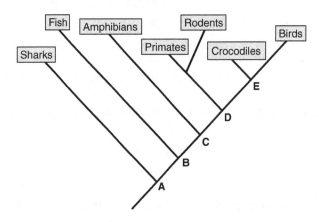

0 1 . 6 Name the feature that is used to classify organisms as vertebrates. **[1 mark]**

_____

0 1 . 7 Which letter represents the most recent common ancestor of amphibians and rodents?

[1 mark]

_____

0 2 This question is about hormones.

0 2 . 1 Thyroxine is a hormone.

Cerys does not produce enough thyroxine. She gains weight easily.

Explain why. **[2 marks]**

_____

_____

Figure 3 shows how production of thyroxine is controlled.

Figure 3

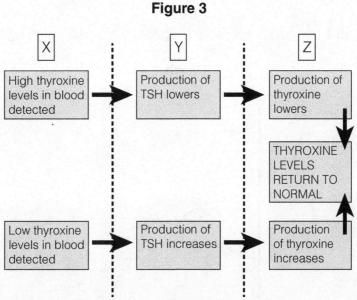

0 2 • 2 Which letter represents the process that the pituitary gland is responsible for?

_____ **[1 mark]**

0 2 • 3 Where in the body is thyroxine produced? **[1 mark]**

_____

0 2 • 4 Cerys goes to the GP to get some test results. She feels anxious. Her heartbeat increases. Cerys is producing the 'flight or fight' hormone.

What is the name of this hormone? **[1 mark]**

_____

0 2 • 5 Hormones produced by the pancreas are responsible for controlling blood glucose levels.

Explain how blood glucose levels are controlled with reference to these hormones and the mechanism of negative feedback. **[6 marks]**

_____

_____

_____

_____

_____

_____

_____

0 2 · 6 **Graph 1** shows how the blood glucose levels of a person without diabetes change shortly after eating a meal.

Draw on **Graph 1** the results you would expect for a person with Type 1 diabetes. **[2 marks]**

Graph 1

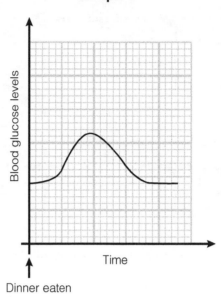

Blood glucose levels

Time

Dinner eaten

0 2 · 7 What is the normal treatment for someone who has Type 1 diabetes? **[1 mark]**

_____

0 3 **Figure 4** shows part of a food web in an area of grassland.

Figure 4

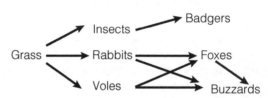

0 3 · 1 Name a top predator. **[1 mark]**

_____

0 3 · 2 What name is given to the grass at the start of the food chain? **[1 mark]**

_____

0 3 · 3 Construct and label a pyramid of biomass for this food chain:

Grass   ⟶   Rabbit   ⟶   Fox **[1 mark]**

0 3 • 4 Myxomatosis is a viral disease that affects rabbits. Death usually occurs within 12 days of infection.

Explain how the introduction of the myxomatosis virus into the rabbit population could affect the fox population. **[1 mark]**

_____

_____

_____

Figure 5 represents the energy flow through the food web in the grassland in one year.

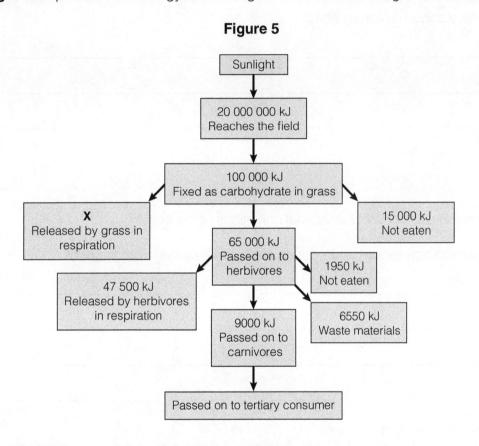

Figure 5

Use **Figure 5** to answer the following questions.

0 3 • 5 Calculate the amount of energy that is released by the grass through respiration (label **X**).

[1 mark]

_____

0 3 • 6 Calculate the percentage of energy that is received by herbivores and passed on to carnivores.

Give your answer to 2 significant figures. **[2 marks]**

_____

| 0 | 3 | • | 7 | How much energy would you expect to be passed from the carnivores to the tertiary consumers? [1 mark]

Tick **one** box.

1000 kJ ☐

2500 kJ ☐

5000 kJ ☐

10 000 kJ ☐

| 0 | 3 | • | 8 | Explain why there will always be a larger biomass of secondary consumers compared to tertiary consumers in a habitat. [2 marks]

_____

_____

_____

| 0 | 4 | Homeostasis controls the internal conditions of the body.

| 0 | 4 | • | 1 | Describe how excess amino acids are safely removed from the body. [4 marks]

_____

_____

_____

_____

| 0 | 4 | • | 2 | The kidneys produce urine.

Which substance is found in the urine of a healthy person? [1 mark]

Tick **one** box.

Glucose ☐

Proteins ☐

Mineral ions ☐

Calcium ☐

0 4 • 3 Explain the role of the hormone ADH in controlling the water content of the body. **[6 marks]**

_____

_____

_____

_____

_____

_____

_____

_____

Table 1 shows the amount of water lost by a healthy person on a hot day.

Table 1

How water is lost	Volume of water lost in cm^3
In urine	900
Through skin	700
Breathed out	300
In faeces	100

0 4 • 4 Calculate the percentage of water lost through sweating. **[1 mark]**

_____

0 4 • 5 Explain how sweating helps to cool the body. **[2 marks]**

_____

_____

_____

0 4 • 6 State **one** other way the skin responds to high temperatures in order to cool the body. **[1 mark]**

_____

0 4 • 7 Explain why maintaining a constant body temperature is important. **[1 mark]**

_____

0 5 Cystic fibrosis is a genetic disorder that is caused by a recessive allele.

0 5 . 1 Explain with the aid of a diagram how two parents without the disorder may produce a child with the disorder. **[3 marks]**

0 5 . 2 Ben and Mia already have one child with cystic fibrosis.

They do not wish to risk having another child with the disorder.

They are told that embryo screening could be used with IVF to help them.

Suggest why some people do not agree with embryo screening. **[1 mark]**

_____

_____

_____

Mendel was a scientist who carried out thousands of experiments on inheritance in plants.

Table 2 shows some of his results.

Table 2

Parent plant 1 Breeding true	Parent plant 2 Breeding true	F1 generation	F2 generation	Ratio of offspring in F2 generation
Round seed	Wrinkled seed	Round	5400 round 1820 wrinkled	2.97 : 1
Red flowers	White flowers	Red	700 red 230 white	3.04 : 1
Tall plants	Small plants	Tall	1800 tall 585 small	

The F1 generation are the results obtained when parent 1 and parent 2 are crossed.

The F2 generation are the results when two of the F1 generation are crossed.

0 5 • 3 Mendel used the term 'breeding true' in his experiments.

What term is used to describe the genotype of an organism that 'breeds true'? **[1 mark]**

_____

0 5 • 4 Calculate the ratio of tall to small plants in the F2 generation.

Give your answer to 2 decimal places. **[1 mark]**

_____

0 5 • 5 Mendel suggested that 'units' controlled characteristics such as flower colour and height in plants, and that these 'units' were inherited.

Suggest why the importance of his work was not understood at that time. **[1 mark]**

_____

0 6 **Figure 6** shows the human brain.

Figure 6

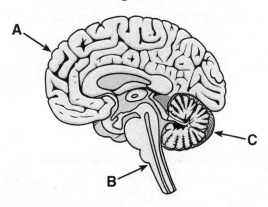

Cerebellar astrocytomas are tumours that develop and grow in the area of the brain called the cerebellum. They usually grow very slowly and do not usually spread. They tend to occur in very young children.

Symptoms may include headaches, vomiting, confusion and blurred vision.

The tumours can be diagnosed in a number of ways and treatment can include surgery or radiotherapy. Often, however, doctors choose to simply keep the tumour under regular observation.

0 6 • 1 Which label on the diagram shows where these tumours grow? **[1 mark]**

_____

0 6 • 2 Name **one** function of this area. **[1 mark]**

_____

0 6 . 3 Once symptoms have been noticed, the tumour can be diagnosed. Suggest **one**
method that could be used for this diagnosis. **[1 mark]**

_____

0 6 . 4 Doctors need to evaluate the risks and benefits of procedures when making decisions about
treatment. Suggest why doctors often choose not to treat these tumours when they
are diagnosed. **[4 marks]**

_____

_____

_____

_____

0 6 . 5 Name the area of the brain that controls automatic actions such as heartbeat
and breathing. **[1 mark]**

_____

0 7 **Figure 7** shows how the sperm and egg cells are formed in humans by meiosis.

Figure 7

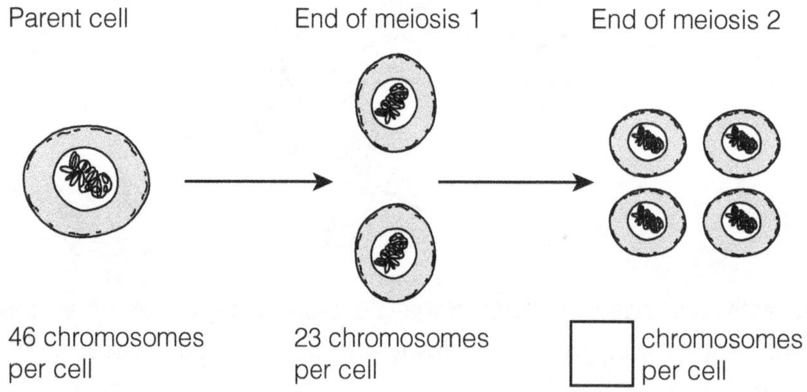

Parent cell	End of meiosis 1	End of meiosis 2
46 chromosomes per cell	23 chromosomes per cell	☐ chromosomes per cell

0 7 . 1 On **Figure 7**, write how many chromosomes will be present in each cell at the end of
meiosis 2. **[1 mark]**

0 7 . 2 Name the organ where meiosis occurs to produce the male gametes. **[1 mark]**

_____

0 7 . 3 If a male gamete carrying an X chromosome fuses with an egg cell, what will be the
sex of the offspring? Explain your answer. **[2 marks]**

_____

_____

0 7 • 4 In asexual reproduction, a different type of cell division is involved.

Name the type of cell division involved in asexual reproduction. **[1 mark]**

_____

0 7 • 5 Suggest **one** advantage and **one** disadvantage of asexual reproduction compared to sexual reproduction. **[2 marks]**

_____

_____

Figure 8 shows a strawberry plant.

Figure 8

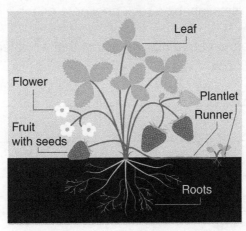

The strawberry plant produces runners. At intervals along a runner, new strawberry plants, called plantlets, grow.

Birds eat the fruit and seeds of the plant. The seeds can be excreted by the birds many miles away. If they land in soil, they will grow into new strawberry plants.

0 7 • 6 Describe how the plants that grow from the runners will be different from the plants that grow from the seeds. **[2 marks]**

_____

_____

_____

_____

Malaria is a serious and sometimes fatal disease caused by a parasite. It is difficult to treat and to date there is no satisfactory vaccination available.

The organism that causes malaria is able to reproduce both sexually and asexually. It requires both types of reproduction to complete its life cycle. Asexual reproduction happens inside humans and sexual reproduction happens inside the mosquito. The parasite is spread when the mosquito bites someone.

In countries where malaria is common, environmental methods to control the spread of the disease include getting rid of pools of stagnant water, and clearing bushes from around houses and planting lemongrass instead.

0 7 . 7 Suggest why these methods are effective in decreasing the incidence of malaria. **[3 marks]**

_____

_____

_____

_____

0 8 A community of organisms inhabit a rock pool on a beach.

0 8 . 1 Name **two** biotic factors and **two** abiotic factors that will affect the numbers and types of organisms in the rock pool. **[2 marks]**

Biotic: _____

_____

Abiotic: _____

_____

In the rock pool there are plant plankton and shrimp.

Graph 2 shows the changing numbers of both throughout a year.

Graph 2

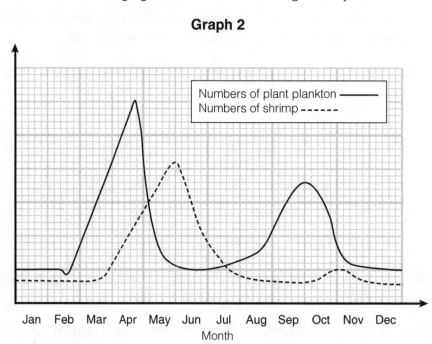

| 0 | 8 | • | 2 | Explain why the numbers of each organism change throughout the year. [4 marks]

_____

_____

_____

_____

_____

_____

_____

_____

Scientists studying the diversity of organisms in rock pools recorded the information shown in **Table 3**.

Table 3

Maximum depth of rock pool at low tide in cm	Average number of different organisms found
20	3
40	3
60	7
80	11
100	14

| 0 | 8 | • | 3 | Describe the pattern in the results. [1 mark]

_____

| 0 | 8 | • | 4 | Suggest a reason for this pattern. [1 mark]

_____

_____

0 9 Carla likes growing flowers for her garden. She buys a packet of sunflower seeds, puts them in a tray of soil, waters them and puts them in a warm place.

Figure 9a shows the seedlings after two weeks.

Figure 9a

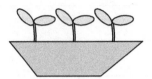

0 9 • 1 Name the plant hormone that is responsible for ending seed dormancy and initiating germination.   **[1 mark]**

_____

Carla moves the seedlings to a window ledge. **Figure 9b** shows the seedlings a week later.

Figure 9b

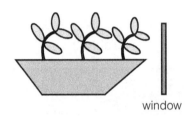

window

0 9 • 2 Explain the role of plant hormones in the seedling's response to light.   **[4 marks]**

_____

_____

_____

_____

Plant hormones can be used in weedkillers.

Farmers are often only concerned about particular species of weeds that compete with their crops. These weeds are called target weeds. Slower-growing weeds that are not a risk to the crops are called non-target weeds.

Spraying of fields with weedkiller, however, destroys both target and non-target weeds.

0 9 • 3 Conservationists believe that the use of weedkillers is reducing the biodiversity of insects and farmland birds. Explain how this could be true.   **[2 marks]**

_____

_____

_____

Fruit that is imported into the country may be treated with plant hormones either before or after arrival.

| 0 | 9 | • | 4 | Give **one** reason for treating fruits with hormones. [1 mark]

_____

| 1 | 0 | Humans interact with the ecosystem in a number of ways.

Almost a fifth of the Amazon rainforest has been destroyed in the past 50 years to make way for huge cattle ranches.

| 1 | 0 | • | 1 | Give **one** reason, other than ranching cattle, why humans carry out large-scale deforestation in tropical areas. [1 mark]

_____

Pollution is a problem caused by the growing population.

Figure 10 is a pie chart that shows the different types of pollutant entering the oceans.

Figure 10

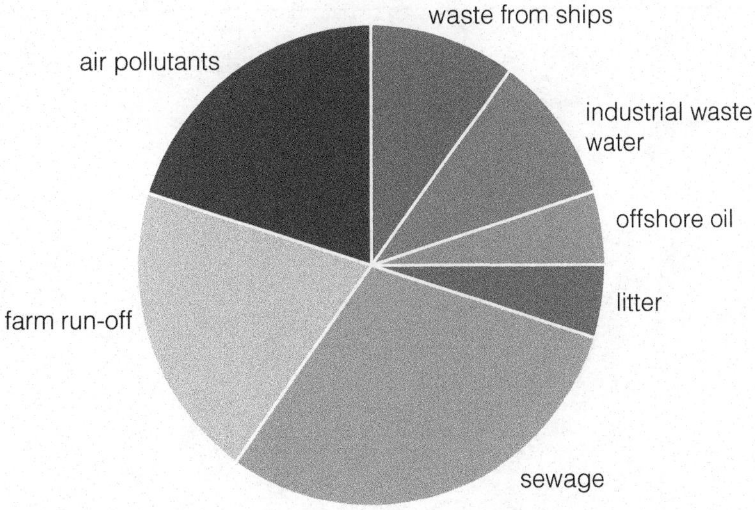

| 1 | 0 | • | 2 | Waste from ships and industrial waste water each count for 10% of the pollutants entering the ocean.

Estimate the percentage of pollution in the ocean caused by sewage. [1 mark]

_____

Increasing levels of pollution in the sea have resulted in a decline in the numbers of many fish – this includes species that are eaten by humans. To prevent numbers falling to levels where there are insufficient adult fish to breed successfully, governments have signed up to sustainable fishing agreements.

| 1 | 0 | • | 3 | Give **one** example of sustainable fishing methods. [1 mark]

_____

Global warming has become a worldwide issue over the past few decades.

| 1 | 0 | • | 4 | Describe the causes and biological consequences of global warming. [6 marks]

_____

_____

_____

_____

_____

_____

_____

END OF QUESTIONS

GCSE (9–1)
Biology
Paper 1H

Higher tier

Time: 1 hour 45 minutes

Materials

For this paper you must have:
- a ruler
- a calculator.

Instructions

- Answer **all** questions in the spaces provided.
- Do all rough work in this book. Cross through any work you do not want to be marked.

Information

- There are 100 marks available on this paper.
- The marks for questions are shown in brackets.
- You are expected to use a calculator where appropriate.
- You are reminded of the need for good English and clear presentation in your answers.
- When answering questions 02.6, 04.4, 08.5 and 10.3 you need to make sure that your answer:
  - is clear, logical, sensibly structured
  - fully meets the requirements of the question
  - shows that each separate point or step supports the overall answer.

Advice

In all calculations, show clearly how you work out your answer.

Name: _____

0 1 **Figure 1** shows a plant cell.

Figure 1

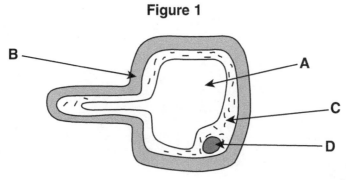

0 1 • 1 State the name of this cell and its function. **[2 marks]**

_____

_____

0 1 • 2 What is the name of the structure labelled B? **[1 mark]**

_____

0 1 • 3 Which label shows where mitochondria will be found? **[1 mark]**

_____

0 1 • 4 The cell does not contain chloroplasts. Explain why. **[2 marks]**

_____

_____

_____

Figure 2 is an image obtained with an electron microscope and shows the surface of a leaf magnified 800 times.

Figure 2

0 1 . 5 Explain how electron microscopy has helped to develop increased knowledge and understanding of cells.

[4 marks]

_____

_____

_____

_____

_____

_____

0 2 Water moves across membranes by osmosis.

0 2 . 1 What is osmosis?

[2 marks]

_____

_____

Some students set up an experiment to investigate the effect of sucrose solution on osmosis. They put potato cubes into different concentrations of sucrose solution and left them for two hours. They weighed the potato cubes at the start and at the end of the experiment. The results are recorded in **Table 1**.

Table 1

	Concentration of sucrose solution					
	0	0.2%	0.4%	0.6%	0.8%	1.0%
Mass of potato at start (g)	12.2	10.5	12.1	15.0	13.5	12.0
Mass of potato after two hours (g)	15.4	12.8	11.6	12.0	11.5	9.2
Change in mass	+3.2	+2.3	−0.5			

0 2 . 2 Complete **Table 1** to show the change in mass for the solutions at 0.6%, 0.8% and 1.0%.

[2 marks]

0 2 . 3 One of the students suggested that calculating the percentage change in mass would give more accurate results. Do you agree? Give a reason for your answer. **[2 marks]**

_____

_____

0 2 . 4 Calculate the percentage change in mass for the potato in 0.6% sucrose solution. **[2 marks]**

_____

0 2 . 5 What can you deduce about the concentration of the solution inside the potato cells before they were placed in the different concentrations of sucrose solution? **[1 mark]**

_____

_____

0 2 . 6 Active transport and diffusion are both processes that allow substances to move into and out of cells.

Describe the differences between active transport and diffusion, and give an example of each process. **[4 marks]**

_____

_____

_____

_____

_____

_____

0 3 **Figure 3** shows the structure of the heart.

Figure 3

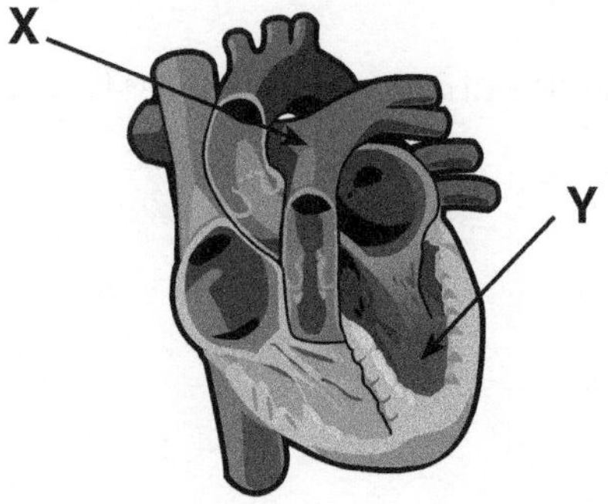

| 0 | 3 | • | 1 | Which statement describes blood vessel X?

[1 mark]

Tick **one** box only.

It is an artery and carries oxygenated blood. ☐

It is an artery and carries deoxygenated blood. ☐

It is a vein and carries deoxygenated blood. ☐

It is a vein and carries oxygenated blood. ☐

| 0 | 3 | • | 2 | Explain how the structure labelled Y is adapted to its function.

[2 marks]

_____

_____

| 0 | 3 | • | 3 | Electrical impulses from the heart muscle cause your heart to beat. The heart rate is controlled by the sinoatrial (SA) node, located in the right atrium. Damage to the SA node can mean that the heart beats too slowly, there may be long pauses between beats, or it switches between slow and fast rhythms.

Suggest what the recommended treatment for such a condition might be.

[1 mark]

_____

The rate of blood flow through a blood vessel can be calculated using this formula:

$$\text{Blood flow} = \frac{\text{Pressure difference}}{\text{Resistance}}$$

Blood vessel A has a pressure difference of 4.

Blood vessel B has a pressure difference of 100.

| 0 | 3 | • | 4 | Assuming a constant blood flow of 5 litres/minute, calculate the resistance in both blood vessels.

[2 marks]

Blood vessel A: _____ mm Hg   Blood vessel B: _____ mm Hg

| 0 | 3 | • | 5 | Which blood vessel is most likely to be an artery? Explain your answer.

[1 mark]

_____

_____

Figure 4 shows blood that has been stained and viewed using a light microscope.

Figure 4

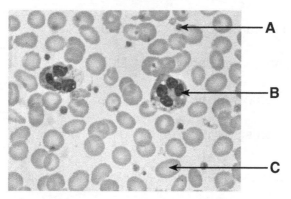

| 0 | 3 | • | 6 | Which letter shows a platelet? [1 mark]

_____

| 0 | 3 | • | 7 | State the function of platelets. [1 mark]

_____

| 0 | 4 | Figure 5 shows a bacterial cell.

Figure 5

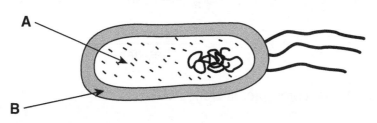

| 0 | 4 | • | 1 | What are the names of the parts labelled A and B? [2 marks]

A _____ B _____

| 0 | 4 | • | 2 | Gonorrhoea is caused by a bacterium. How is gonorrhoea transmitted? [1 mark]

_____

| 0 | 4 | • | 3 | Scientists can grow bacteria in petri dishes on nutrient agar. Before the agar is placed into the petri dish, it is heated to 121°C for 20 minutes. Explain why. [1 mark]

_____

| 0 | 4 | • | 4 | *Salmonella* food poisoning is a disease caused by bacteria. **Figure 6** shows some apparatus. Describe how scientists could use the apparatus to find out if a sample of chicken broth contained *Salmonella* bacteria. **[4 marks]**

Figure 6

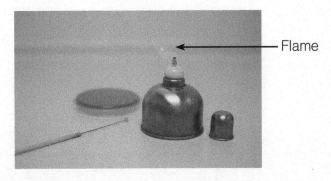

Flame

_____

_____

_____

_____

The mean division time of the *Salmonella* bacterium that causes food poisoning is 20 minutes. **Table 2** shows how the numbers of these bacteria change over a period of two hours when the chicken broth is left at room temperature.

Table 2

Time in minutes	Number of bacteria per cm³
0	100
20	200
40	400
60	800
80	1600
100	3200
120	

| 0 | 4 | • | 5 | In **Table 2**, fill in the missing number of bacteria at 120 minutes. **[1 mark]**

A healthy person would normally need to ingest 1×10^5 bacteria before they become ill with food poisoning.

Use **Table 2** to help answer the following question.

0 4 . 6 A healthy person consumes 100 cm³ of chicken broth that has been left at room temperature for 80 minutes. Are they likely to suffer from food poisoning?

Explain your answer. **[2 marks]**

_____

_____

_____

0 4 . 7 Symptoms of *Salmonella* food poisoning include diarrhoea and stomach cramps.

Name the type of substance produced by the *Salmonella* bacterium that is responsible for these symptoms. **[1 mark]**

_____

0 5 In March 2017, researchers announced they had cured a 13-year-old boy who had sickle cell disease. The disease is caused by a faulty gene that causes abnormal haemoglobin to be produced. The abnormal haemoglobin causes the red blood cells to change shape and clump together.

Researchers removed some stem cells from the boy and inserted an extra gene into the stem cells to interfere with the faulty haemoglobin gene. The researchers then put the stem cells back into the boy's body. Within three months the boy had started to produce normal haemoglobin.

0 5 . 1 State the function of haemoglobin. **[1 mark]**

_____

0 5 . 2 The genes coding for haemoglobin are found in which cellular structure? **[1 mark]**

_____

0 5 . 3 What is a stem cell? **[1 mark]**

_____

0 5 . 4 From which part of the boy's body is it likely that researchers took the stem cells? **[1 mark]**

Tick **one** box.

Blood ☐

Brain ☐

Bone marrow ☐

Skin ☐

| 0 | 5 | • | 5 | Stem cells from human embryos can be used to treat a number of conditions.

Suggest **one** possible risk of using stem cells. **[1 mark]**

_____

Another disease that affects red blood cells is autoimmune haemolytic anaemia. People with this condition have an overactive immune system that destroys the body's own red blood cells, causing anaemia. A drug company has recently developed a new drug (_Haematodome_) to treat the condition. The drug works because it suppresses the immune system.

| 0 | 5 | • | 6 | Describe the development and testing processes that must happen before _Haematodome_ can be made available for public use. **[4 marks]**

_____

_____

_____

_____

_____

| 0 | 6 | Charlie is undertaking a project on respiration.

| 0 | 6 | • | 1 | Complete the symbol equation for respiration: **[2 marks]**

$6O_2$ + _____ $\longrightarrow$ $6CO_2$ + _____

Charlie wants to investigate how his breathing rate changes with exercise of different intensities.

He measures his breathing rate at rest, after walking briskly for two minutes, jogging for two minutes and sprinting for two minutes. He records his results in **Table 3**.

Table 3

Exercise	Breaths per minute
Resting (no exercise)	16
Walking	18
Jogging	25
Sprinting	32

0 6 • 2 Explain why Charlie's breathing rate increases as the intensity of exercise increases. **[3 marks]**

_____

_____

_____

0 6 • 3 Describe another way Charlie's breathing will change as exercise intensity increases. **[1 mark]**

_____

When Charlie sprints, he begins to respire anaerobically. Lactic acid, produced by anaerobic respiration, will build up in his muscle cells.

0 6 • 4 Explain what happens to the lactic acid. **[3 marks]**

_____

_____

Charlie wanted to know what mechanism in the body was responsible for controlling breathing rate. He searched the internet and found **Graphs 1** and **2** below.

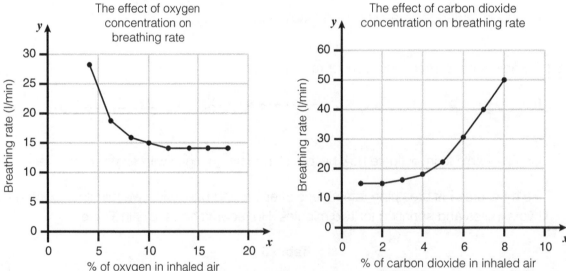

0 6 • 5 Which gas has the most effect on breathing rate?

Use evidence from **Graph 1** and / or **Graph 2** to support your answer. **[2 marks]**

_____

_____

_____

Look at **Figure 7**. Fish have gills to remove dissolved oxygen from water. Each gill filament has hundreds of tiny protrusions called lamellae. Blood runs through the lamellae whilst water passes across them.

Figure 7

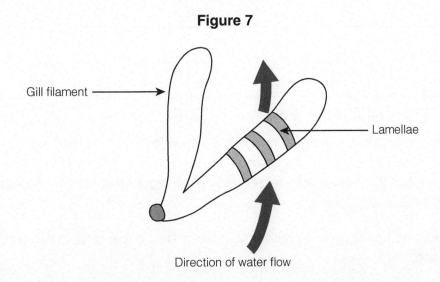

Gill filament

Lamellae

Direction of water flow

0 6 . 6 Suggest how the lamellae help to maximise diffusion of oxygen from the water to the blood of the fish. **[1 mark]**

_____

0 7 Whooping cough is a highly contagious bacterial infection of the lungs and airways.

0 7 . 1 Describe how cells in the trachea and bronchi are adapted to prevent pathogens reaching the lungs. **[2 marks]**

_____

_____

_____

If the bacteria causing whooping cough enters the body, white blood cells respond to the pathogen in a number of ways, one of which is shown in **Figure 8**.

Figure 8

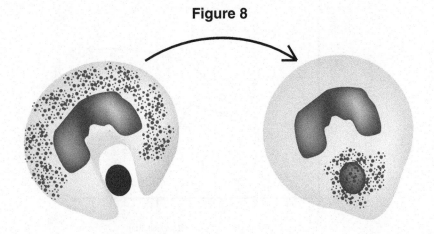

$\boxed{0 \; 7} \cdot \boxed{2}$ Name and describe the process shown in **Figure 8**. **[3 marks]**

_____

_____

_____

_____

The first symptoms of whooping cough are similar to those of a cold, such as a runny nose, red and watery eyes, a sore throat and a slightly raised temperature. People diagnosed during the first three weeks of infection may be prescribed antibiotics and are also told to take painkillers such as aspirin.

$\boxed{0 \; 7} \cdot \boxed{3}$ Antibiotics and painkillers are both drugs. Describe the role of each in the treatment of disease. **[2 marks]**

_____

_____

$\boxed{0 \; 8}$ This question is about photosynthesis.

$\boxed{0 \; 8} \cdot \boxed{1}$ Write the word equation for photosynthesis. **[2 marks]**

_____ + _____ ⟶ _____ + _____

Some students measured the rate of photosynthesis of pond weed at different light intensities in high and low concentrations of carbon dioxide.

Graph 3 shows their results.

Graph 3

X — X   Low concentration of carbon dioxide

O — O   High concentration of carbon dioxide

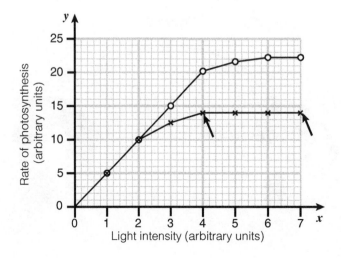

| 0 | 8 | • | 2 | Explain why the rate of photosynthesis has stopped increasing between the two arrows in **Graph 3**. [1 mark]

_____

| 0 | 8 | • | 3 | At a light intensity of 4 arbitrary units, how much faster is the rate of photosynthesis of the pond weed in high carbon dioxide concentrations compared to low carbon dioxide concentrations? [1 mark]

_____ arbitrary units

| 0 | 8 | • | 4 | The glucose produced in photosynthesis may be stored as starch.

Give **two** other ways the glucose may be used. [2 marks]

_____

_____

Figure 9 shows a cross-section of a leaf.

Figure 9

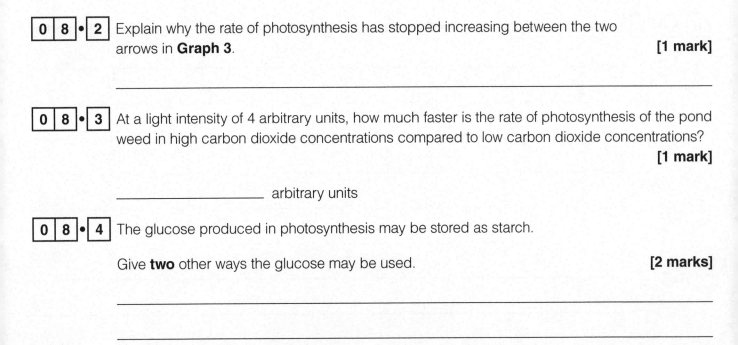
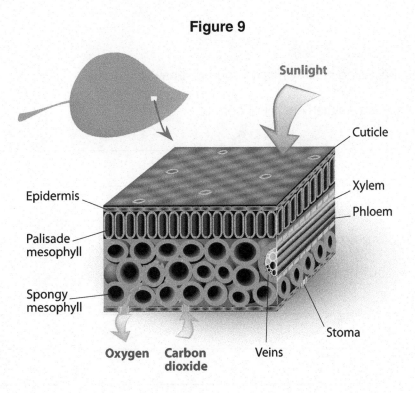

0 8 • 5 Explain how the leaf in **Figure 9** is adapted for photosynthesis. **[6 marks]**

_____

_____

_____

_____

_____

_____

_____

_____

_____

_____

0 9 Petra says that being healthy means you don't have a disease.

0 9 • 1 Explain why Petra's statement is not correct. **[1 mark]**

_____

_____

Cancers form when cells begin to divide uncontrollably.

0 9 • 2 Explain the difference between benign and malignant tumours. **[2 marks]**

_____

_____

_____

Sometimes when a person has an illness or disease, they are more at risk from other illnesses or diseases.

| 0 | 9 | • | 3 | What illness or disease is a person who has had human papilloma virus (HPV) more at risk from? **[1 mark]**

Tick **one** box.

Cervical cancer ☐ Cold sores ☐

Allergic reactions ☐ Stomach cancer ☐

Researchers often use scatter graphs to determine if there is a link between two factors in terms of risk.

Researchers investigating links to diet and bowel cancer looked at the average adult weekly meat consumption per head of population in different countries. They plotted this against the incidence of bowel cancer per 100 000 of the country's population. The results are shown in **Graph 4**.

Graph 4

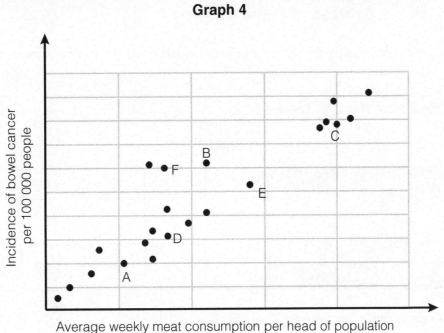

Average weekly meat consumption per head of population

| 0 | 9 | • | 4 | What conclusions could you draw from **Graph 4**? **[1 mark]**

_____

_____

| 0 | 9 | • | 5 | What country has twice the incidence of bowel cancer as the country labelled D? **[1 mark]**

_____

| 0 | 9 | • | 6 | If the average weekly meat consumption per person in country C was 1.6 kg, estimate the average weekly meat consumption per person living in country E. **[1 mark]**

_____

1 0 Some students wanted to compare water uptake by two different plants.

They placed each plant in a measuring cylinder containing water.

They placed a layer of oil on top of the water.

Figure 10 shows the experiment.

Figure 10

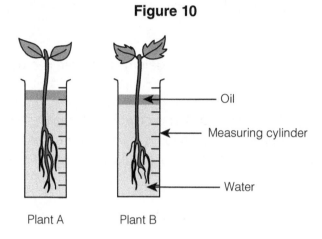

Plant A   Plant B

The students measured the amount of water in each measuring cylinder every day for five days.

Their results are shown in **Table 4**.

Table 4

Time in days	0	1	2	3	4	5
Volume of water in cylinder of Plant A in cm³	40	36	33	31	30	28
Volume of water in cylinder of Plant B in cm³	38	33	27	24	21	18

The average water loss per day for plant A was 2.4 cm³.

1 0 · 1 Calculate the average water loss per day for plant B. **[1 mark]**

_____

_____

| 1 | 0 | • | 2 | Suggest **two** variables the students would need to control to obtain a fair comparison between the two plants. **[2 marks]**

_____

_____

Water vapour is lost through the stomata.

Figure 11 shows a guard cell.

Figure 11

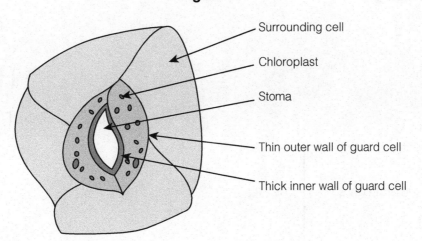

| 1 | 0 | • | 3 | Explain how guard cells control the opening and closing of stomata. **[6 marks]**

_____

_____

_____

_____

_____

_____

_____

END OF QUESTIONS

GCSE (9–1)
Biology
Paper 2H

Higher tier

Time: 1 hour 45 minutes

Materials

For this paper you must have:
- a ruler
- a calculator.

Instructions

- Answer **all** questions in the spaces provided.
- Do all rough work in this book. Cross through any work you do not want to be marked.

Information

- There are 100 marks available on this paper.
- The marks for questions are shown in brackets.
- You are expected to use a calculator where appropriate.
- You are reminded of the need for good English and clear presentation in your answers.
- When answering questions 02.5, 05.4, 07.5 and 09.5 you need to make sure that your answer:
  - is clear, logical, sensibly structured
  - fully meets the requirements of the question
  - shows that each separate point or step supports the overall answer.

Advice

In all calculations, show clearly how you work out your answer.

Name: _____

0 1   **Figure 1** is part of a food web from a small stretch of a canal.

Figure 1

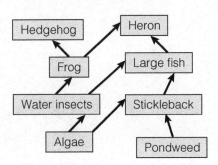

0 1 • 1   What trophic level is occupied by the frog?   **[1 mark]**

_____

Waste from a local factory is discharged into the canal. This reduces the pH of the water, which results in female sticklebacks being unable to produce eggs.

0 1 • 2   What effect will this have on the number of frogs and insects over a period of time?   **[2 marks]**

_____

_____

_____

A local wildlife group want to know more about the numbers of hedgehogs along the stretch of canal.

They set some traps along the footpath one evening. The following morning they mark the six hedgehogs they have caught with a special dye before releasing them.

A week later, the group return to the same area and set more traps overnight. The following morning they find nine hedgehogs, three of which are marked with the special dye.

0 1 • 3   Use the formula to find the number of hedgehogs along the stretch of canal.   **[2 marks]**

$$\frac{\text{Number in first sample} \times \text{Number in second sample}}{\text{Number of marked organisms in second sample}}$$

_____

_____

0 1 • 4   Suggest why it is difficult to accurately estimate the number of organisms along a stretch of canal.   **[2 marks]**

_____

_____

| 0 | 2 | | **Graph 1** shows how two hormones change during the menstrual cycle.

Graph 1

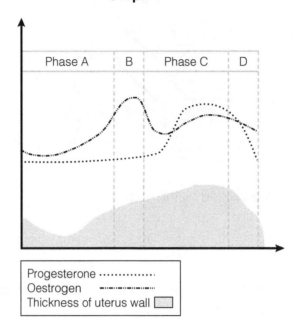

| 0 | 2 |•| 1 | Describe the effect of oestrogen on the thickness of the uterus wall. **[1 mark]**

_____

| 0 | 2 |•| 2 | In phase A the follicle develops and matures.

What happens to the follicle in phase B? **[1 mark]**

_____

| 0 | 2 |•| 3 | Describe the role of progesterone in phase C. **[1 mark]**

_____

The hormones oestrogen and progesterone are both produced by the ovaries.

| 0 | 2 |•| 4 | Name the site of production for follicle stimulating hormone (FSH) and luteinising hormone (LH). **[1 mark]**

_____

Hormones can be used as contraceptives.

0 2 • 5 Discuss the advantages and disadvantages of the oral contraceptive pill when compared with barrier methods of contraception. **[6 marks]**

_____

_____

_____

_____

_____

_____

Hormones can also be used in fertility treatment.

Fatima and Ahmed have been trying to have a baby for several years and have been offered IVF. Fatima will be given hormones to stimulate the maturation of eggs. The eggs will be collected and fertilised with Ahmed's sperm in the laboratory.

0 2 • 6 Describe what will happen next. **[2 marks]**

_____

_____

_____

0 3 **Figure 2** shows the organisms involved in the constant recycling of carbon.

Figure 2

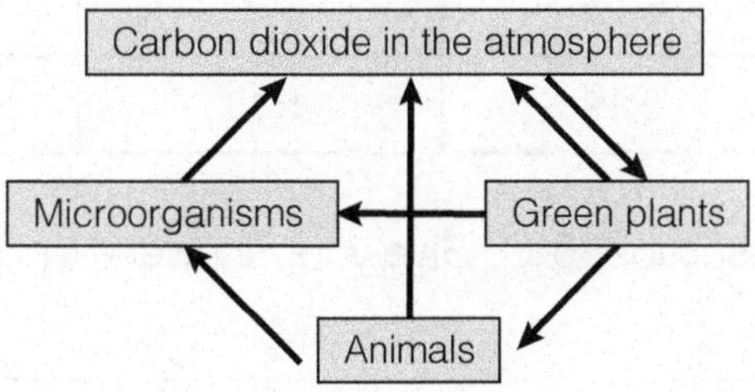

0 3 • 1 Describe the role of microorganisms in this cycle. **[3 marks]**

_____

_____

_____

_____

0 3 . 2 Describe what happens to the carbon dioxide from the atmosphere once it has entered the plants.

[2 marks]

_____

_____

_____

0 4 Some students wanted to investigate the rate of decay of milk kept at different temperatures.

When milk is decayed by bacteria, lactic acid is formed. The students therefore decided that measuring the pH would give an indication of how much the milk had decayed.

They put 10 cm³ of milk into three test tubes, placed a cotton wool bung in the top of each and left the tubes for 24 hours in a fridge at 5°C. They then measured the pH of the milk.

They repeated the test in water baths at 10°C, 15°C, 20°C, 25°C, 30°C and 35°C.

Their results are shown in **Table 1**.

Table 1

| Temperature °C | pH after 24 hours | | | |
	Test tube 1	Test tube 2	Test tube 3	Mean
5	7.71	7.81	7.90	7.8
10	7.71	7.72	7.66	7.7
15	7.54	7.55	7.58	7.6
20	7.39	7.42	7.46	7.4
25	7.08	7.09	7.12	7.1
30	6.24	6.21	6.16	
35	5.19	5.17	5.17	5.2

0 4 . 1 Calculate the mean result for 30°C. Give your answer to 1 decimal place.

[2 marks]

_____

_____

0 4 · 2 Plot the mean results on the axes in **Graph 2**. Draw a line or curve. **[3 marks]**

Graph 2

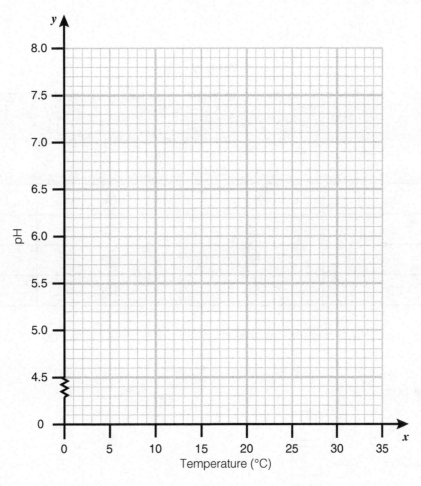

0 4 · 3 Describe the effect of temperature on the decay of milk. **[2 marks]**

_____

_____

_____

Anaerobic decay produces methane gas.

Research was carried out to compare the effect of using different substrates on the amount of methane produced.

The results are shown in **Table 2**.

Table 2

	Cumulative amount of biogas produced (arbitrary units) for each substrate		
Time in days	100% cow manure	50% pig manure 50% cow manure	100% pig manure
5	15	36	28
10	35	65	51
15	65	89	64
20	80	100	72

0 4 • 4 Between which days was the maximum amount of biogas produced from the 100% cow manure? **[1 mark]**

Tick **one** box.

0–5 days ☐

6–10 days ☐

11–15 days ☐

16–20 days ☐

0 4 • 5 By how much was the percentage yield of biogas increased after 20 days by using a mixture of 50% cow manure and 50% pig manure compared to cow manure alone? **[1 mark]**

_____%

0 4 • 6 If only one type of manure was available, would you recommend cow or pig manure?

Using only data from **Table 2**, give a reason for your answer. **[1 mark]**

_____

_____

_____

0 5 The nervous system enables humans to react to their surroundings.

Information is passed from receptors to the central nervous system, which then coordinates a response by an effector.

0 5 • 1 Name **two** receptors found in the skin. **[2 marks]**

_____ and _____

0 5 • 2 Motor neurones carry impulses to muscles and endocrine glands.

Describe how each of them responds on receiving an impulse. **[2 marks]**

_____

_____

Figure 3 shows a synapse.

Figure 3

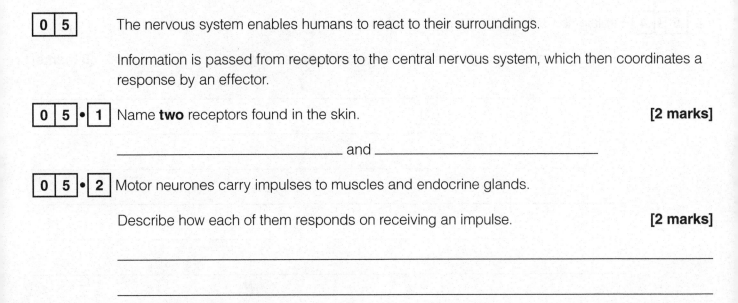

Information passes from one neurone to another via synapses.

0 5 • 3 Draw an arrow on **Figure 3** to show the direction information is travelling. **[1 mark]**

0 5 • 4 Explain how information passes across a synapse.

In your answer include the names of the **two** neurones involved. **[5 marks]**

_____

_____

_____

_____

_____

_____

_____

_____

_____

_____

0 5 • 5 Reflex actions do not involve the conscious part of the brain.

Why are reflex actions important? **[1 mark]**

_____

0 6 DNA is a polymer made of two strands of nucleic acid.

Each strand consists of a chain of nucleotides.

Figure 4 shows a nucleotide.

Figure 4

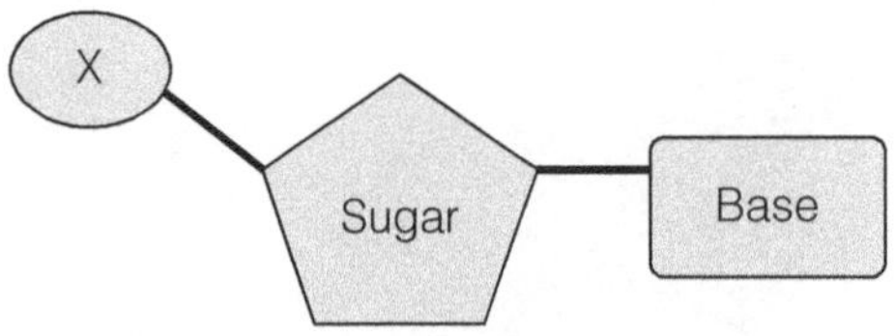

0 6 . 1 What is the molecule labelled X? **[1 mark]**

Tick **one** box.

A nitrate ☐

A carbonate ☐

A phosphate ☐

An acetate ☐

The bases on one strand of nucleic acid form cross links with bases on the other strand. The four bases are recognised by the first letters of their names.

Here is the sequence of bases on one strand of nucleic acid.

A T T G C T G C C C A T

0 6 . 2 Write the letters for the base sequence on the complementary strand of DNA. **[1 mark]**

___ ___ ___ ___ ___ ___ ___ ___ ___ ___ ___ ___

0 6 . 3 In the sequence above, how many amino acids are coded for? **[1 mark]**

_____

0 6 . 4 Describe the role of mRNA in protein synthesis. **[3 marks]**

_____

_____

_____

0 6 . 5 All cells contain the gene for haemoglobin production, yet only red blood cells produce haemoglobin. Explain why. **[1 mark]**

_____

0 6 . 6 Occasionally an error occurs during the copying process and the wrong nucleotide is inserted into the mRNA. Describe how this might affect the protein that is synthesised. **[1 mark]**

_____

_____

Scientists spent more than ten years studying the entire human genome.

0 6 . 7 Give **one** reason why scientists want to understand the human genome. **[1 mark]**

_____

_____

0 7   Pigs produce manure that contains high amounts of phytate, a phosphorus-containing chemical. When farmers use pig manure as fertiliser, the phytate gets washed from fields into streams, rivers and lakes, which results in the death of many organisms that inhabit them.

Figure 5

The traditional method of reducing the amount of phytate in pig manure is by adding the enzyme phytase to pig food to help digest the phytate. This is quite costly.

Scientists have recently created a genetically engineered pig which they call 'enviropig'. The pig has been engineered to produce the enzyme phytase in its salivary glands so it can better digest phytate. **Figure 6** shows the stages in the engineering process.

Figure 6

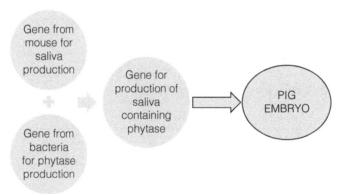

0 7 . 1   Describe the main steps in creating a genetically engineered organism.   **[3 marks]**

_____

_____

_____

_____

0 7 . 2   The 'enviropig' is described as environmentally friendly. Suggest why.   **[1 mark]**

_____

_____

0 7 . 3   Many consumers are against the idea of the 'enviropig'. Suggest why.   **[1 mark]**

_____

_____

0 7 • 4 Suggest why many farmers are in favour of the 'enviropig'. **[1 mark]**

_____

_____

Over 30 different engineered 'enviropig lines' have been produced by inserting the mouse and bacterial genes into different chromosome locations in the pig. Breeding these pigs for further research is a slow and costly procedure.

Adult cell cloning would provide numerous clones of pigs for further research.

0 7 • 5 Describe how scientists could clone the 'enviropig'. **[5 marks]**

_____

_____

_____

_____

_____

_____

0 8 In 2016, two families on holiday in Australia discovered a series of fossils thought to be from _Australopachycormus hurleyi,_ a three-metre long swordfish-like predator that used its sword to injure competitors and kill prey.

0 8 • 1 What is a fossil? **[2 marks]**

_____

_____

0 8 • 2 _Australopachycormus hurleyi_ is now extinct.

Give **one** environmental and **one** biological reason why animals may become extinct.

[2 marks]

_____

_____

Figure 7 shows a modern-day swordfish.

Figure 7

Modern-day swordfish with very long swords have evolved over time from swordfish with shorter swords.

Darwin and Wallace were scientists with ideas about evolution.

| 0 | 8 | • | 3 | Use their ideas to explain the change in sword length. **[4 marks]**

_____

_____

_____

_____

_____

_____

_____

_____

_____

_____

Darwin's ideas about evolution and natural selection were met with controversy when he first published them.

| 0 | 8 | • | 4 | Explain why people did not believe Darwin's theory. **[2 marks]**

_____

_____

_____

| 0 | 9 | Food security is having enough food to feed a population.

Many countries in the world do not have food security.

| 0 | 9 |•| 1 | Give **two** biological reasons why some countries do not have food security.   **[1 mark]**

_____

_____

Figures 8 and **9** show two different methods of farming cows.

Figure 8

Figure 9

| 0 | 9 |•| 2 | Discuss the commercial advantages of each method.   **[4 marks]**

_____

_____

_____

_____

_____

_____

Tom is a farmer. Until a few years ago, he used to grow a variety of different vegetables in his fields. Today he grows only russet potatoes.

Tom says that growing a single crop makes it easier and therefore more efficient to both plant and harvest the crop.

The local conservation group believe that growing only one type of crop reduces biodiversity.

0 9 · 3 Explain why growing only one type of crop reduces biodiversity. **[2 marks]**

_____

_____

_____

0 9 · 4 What could the conservation group ask Tom to do in order to reduce the impact of growing a single crop on the biodiversity of the area? **[1 mark]**

_____

Figure 10 shows a zebu cow.

Figure 10

In Africa, smallholder farmers use zebu cattle. These cattle can withstand high temperatures and survive long periods of food and water shortages. However, they produce only a few litres of milk each day.

Taurine cattle, which live in cooler climates, can produce 20 litres of milk a day.

0 9 · 5 Explain how farmers could use selective breeding to produce cattle that are tolerant to heat and water shortages, and also produce good milk yields. **[4 marks]**

_____

_____

_____

_____

_____

_____

Mycoprotein is a high protein food produced from the fungus *Fusarium*.

The fungus is grown inside fermenters for up to six weeks. It doubles its biomass every six hours. At the end of six weeks it is harvested, purified and dried.

0 9 • 6 Suggest **one** advantage and **one** disadvantage of using mycoprotein to meet the increasing demands for food in developing countries. **[2 marks]**

Advantage: _____

Disadvantage: _____

1 0 **Figure 11** shows the structure of the eye.

Figure 11

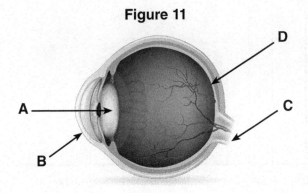

1 0 • 1 Which letter represents the cornea? **[1 mark]**

_____

1 0 • 2 Describe how the pupil responds to dim light. **[1 mark]**

_____

The function of the lens is to focus the light.

1 0 • 3 Which letter on **Figure 11** represents the lens? **[1 mark]**

_____

1 0 • 4 Which letter on **Figure 11** shows where the light is focused? **[1 mark]**

_____

A person with healthy eyesight is able to focus on both near and far objects with ease. This is called accommodation.

1 0 • 5 Describe how the shape of the lens changes when focusing on near and distant objects. **[2 marks]**

_____

_____

| 1 | 1 | Bacteria can evolve quickly because they reproduce at such a fast rate.

Some bacteria have evolved to become resistant to antibiotics.

MRSA is a bacteria that has evolved to become resistant to the antibiotic methicillin.

Graph 3 shows how the number of cases of MRSA blood infections has changed over seven years.

Graph 3

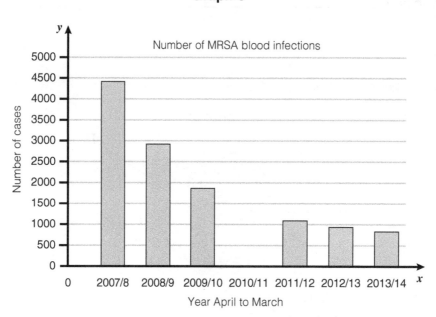

| 1 | 1 | • | 1 | In the year 2010/11 there were 1480 cases of MRSA blood infections.

Plot this result on **Graph 3**. [1 mark]

In hospitals, improved hygiene has reduced the incidence of antibiotic-resistant organisms.

| 1 | 1 | • | 2 | Calculate the percentage decrease in the number of MRSA infections between 2007/8 and 2013/14.

Give your answer to 1 significant figure. [1 mark]

_____

_____

_____%

| 1 | 1 | • | 3 | Describe how antibiotic-resistant populations of bacteria develop. [3 marks]

_____

_____

_____

END OF QUESTIONS

SET A: PAPER 1

01.1 White blood cells **(1)** recognise the pathogen **(1)** and produce antibodies **(1)** quickly **(1)**.

01.2 $\frac{420}{531} \times 100 = 79.09\%$ **(1)** Percentage = 79% **(1)**

01.3 By air **(1)**

01.4 **Any one from:**
- Sexual contact
- Contact with body fluids / blood
- Sharing needles **(1)**.

01.5 They are caused by viruses that live inside cells **(1)**. Drugs that kill the virus would also damage cells **(1)**. Viruses are not destroyed by antibiotics **(1)**.

TOTAL MARKS FOR QUESTION 1 = 11

02.1 **This is a model answer, which would gain all 6 marks:**
Water enters through the root hair cells, which have a large surface area, by osmosis. It travels up the stem through the xylem, which is composed of long, thin, dead cells that have no end cell walls. It is pulled up the xylem by the transpiration stream. It is lost from the plant by evaporation through the stomata, which can be opened and closed by the guard cells to control the amount of water lost.

02.2 0.9 g / 900 mg **(1)**

02.3 $\frac{900}{180} = 5$ mg/minute **(1)**

02.4 Active transport **(1)**

02.5 Stunted / poor growth **(1)**

TOTAL MARKS FOR QUESTION 2 = 10

03.1 Enzyme A has a certain shape / active site, which substrate X fits into **(1)**. Substrate Y will not fit the enzyme shape / active site **(1)**.

03.2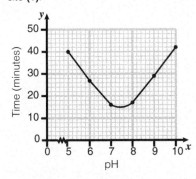

(1 mark for x and y axes correct way round; 1 mark for both axes with suitable scales and labelled; 1 mark for at least five points correctly plotted; 1 mark for joining points with a curve)

03.3 The optimum pH for amylase is between 7.0 and 7.9 (accept any single value given in this range) **(1)**.

03.4 No reaction / time longer than 42 minutes **(1)**

03.5

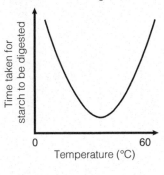

(1)

TOTAL MARKS FOR QUESTION 3 = 9

04.1 Hybridoma **(1)**

04.2 Mouse **(1)**

04.3 The monoclonal antibodies deliver the toxic chemicals directly to the leukaemia cells **(1)**, but healthy cells are not affected **(1)**.

04.4

Scenario	Colour of zone 2	Colour of zone 3
Urine from pregnant woman	Blue	Blue
Urine from non-pregnant woman	Colourless	Blue

(1 mark for each row that is correct)

04.5 Control / to check the test has worked correctly **(1)**

TOTAL MARKS FOR QUESTION 4 = 7

05.1 Purple **(1)**

05.2 The liver **(1)** produces bile **(1)**, which neutralises the acid.

05.3 Lipase digests fats to fatty acids and glycerol **(1)**. Amylase digests starch to maltose / sugars **(1)**. Protease digests proteins to amino acids **(1)**.

05.4 **This is a model answer, which would gain all 6 marks:**
The small intestine is long, which allows time for products to be absorbed. It is covered in villi, which give a large surface area. The lining of the intestine is thin, which means there is a short path for diffusion / active transport. It also has a good blood supply to take the nutrients away once absorbed to maintain concentration gradients.

TOTAL MARKS FOR QUESTION 5 = 12

06.1 **This is a model answer, which would gain all 6 marks:**
Aerobic respiration requires oxygen but anaerobic respiration does not. Aerobic respiration produces carbon dioxide whereas anaerobic respiration produces lactic acid. Aerobic respiration releases more energy than anaerobic respiration because in aerobic respiration glucose is completely broken down whereas anaerobic respiration is incomplete breakdown of glucose. Aerobic respiration takes place in the mitochondria whereas anaerobic respiration takes place in the cytoplasm.

06.2 Glycerol and fatty acids **(1)**

06.3 It is a component of cell walls **(1)**.

06.4 They are broken down to amino acids **(1)**, turned into urea **(1)** and excreted by the kidneys **(1)**.

06.5 Keeping warm **(1)**; movement **(1)** (or any other appropriate answers).

06.6 Fermentation **(1)**

06.7 Ethanol + carbon dioxide **(both products for 1 mark)**

06.8 Fermentation will stop / no ethanol will be produced **(1)** because the enzyme is damaged / denatured **(1)**.

TOTAL MARKS FOR QUESTION 6 = 17

07.1 **This is a model answer, which would gain all 6 marks:**
Prokaryotes are smaller than eukaryotes. Unlike eukaryotes, they do not have a nuclear membrane; the DNA lies free in the cytoplasm. Prokaryotes may contain plasmids whereas eukaryotes do not. Prokaryotes are single-celled whereas eukaryotes are multicellular. An example of a prokaryote is a bacterium. There are many examples of eukaryotes, such as humans, worms, daisies.

07.2 $\frac{5.8}{100} \times 1000 = 58$ µm **(1)**

07.3 Epithelial cells **(1)**

Answers

07.4 The nucleus is dividing **(1)**.

07.5 For growth and repair of tissues **(1)**

07.6 It has a large number of mitochondria **(1)** because these are sites of respiration / where energy is released **(1)**.

TOTAL MARKS FOR QUESTION 7 = 12

08.1 Movement of molecules / particles **(1)** from an area of high concentration to an area of low concentration **(1)**.

08.2 A = carbon dioxide; B = oxygen **(both responses needed for 1 mark)**

08.3 The alveoli provide a large surface area **(1)**, have very thin walls **(1)**, have moist walls **(1)** and an excellent blood supply **(1)**.

08.4 Volume / concentration of glucose solution **(1)**; volume of water **(1)**

08.5 10°C took six minutes for the glucose testing stick to change colour; 20°C was twice as fast at three minutes **(1)**.

08.6 An increase in temperature causes the molecules / particles to move faster **(1)**.

TOTAL MARKS FOR QUESTION 8 = 11

09.1 **Any two from:**
- Stunted growth
- Spotted or decayed leaves
- Growths
- Visible pests
- Discolouration
- Malformed stems or leaves **(2)**.

09.2 **Any one from:**
- Look in a gardening manual
- Take plants to a laboratory
- Use a testing kit **(1)**.

09.3 To make proteins **(1)**

09.4 **This is a model answer, which would gain all 6 marks:**
Plant cells have tough cell walls and leaves have a waxy coat. Both these mechanisms prevent damage to plant tissue, which could then allow pathogens to enter. Should pathogens enter, plants produce antimicrobial substances to kill them. Plants have a number of mechanisms to deter animals from trying to eat them, such as producing poisons, mimicking dangerous organisms and having thorns and hairs. This is important because animals could wound the plant allowing pathogens to enter and could also carry pathogens to the plant.

09.5 Fungus **(1)**

TOTAL MARKS FOR QUESTION 9 = 11

SET A: PAPER 2

01.1 Acts as a lure for prey **(1)**; lights up the area to help it navigate **(1)**
01.2 Analysis of DNA **(1)**
01.3 First part / *Histiophryne* **(1)**
01.4 A group of organisms that can breed to produce fertile offspring **(1)**
01.5 Class **(1)**
01.6 Presence of a backbone **(1)**
01.7 C **(1)**

TOTAL MARKS FOR QUESTION 1 = 8

02.1 Low thyroxine levels cause a low metabolic rate **(1)**. Therefore Cerys will not release energy quickly from the food she eats – any excess will be stored as fat **(1)**.
02.2 Y **(1)**
02.3 Thyroid gland **(1)**
02.4 Adrenaline **(1)**
02.5 **This is a model answer, which would gain all 6 marks:**
When glucose levels rise, the pancreas produces insulin. This causes the liver to convert excess glucose to glycogen for storage. It also causes glucose to move into cells. A return to normal glucose levels is detected by the pancreas and insulin production stops. This is an example of negative feedback. If glucose levels drop, for example during exercise, the pancreas produces glucagon, which causes the liver to convert stored glycogen into glucose and release it into the bloodstream.
02.6

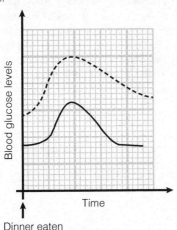

(1 mark for inserting a line above the person without diabetes; 1 mark for a slower decrease in glucose than the person without diabetes)
02.7 Insulin injections **(1)**

TOTAL MARKS FOR QUESTION 2 = 14

03.1 Buzzard / Badger **(1)**
03.2 Producer **(1)**
03.3

 (1)

03.4 Fox population would decrease because there is less food to eat / fox population would increase because buzzard population would decrease due to lack of food **(1)**.
(No mark for increase or decrease – mark awarded for the correct reason)

03.5 65 000 + 15 000 = 80 000
100 000 – 80 000 = 20 000 kJ **(1)**
03.6 $\frac{9000}{65\,000} \times 100$ **(1)** = 14% **(1)**
03.7 1000 kJ **(1)**
03.8 Only a small amount of energy from the secondary consumer is passed to the tertiary consumer **(1)**. So the tertiary consumer needs to eat a large number of secondary consumers to get enough energy to survive **(1)**.

TOTAL MARKS FOR QUESTION 3 = 10

04.1 They are taken to the liver **(1)**, where they are converted to urea **(1)**, which travels in the blood to the kidneys **(1)** and is excreted in urine **(1)**.
04.2 Mineral ions **(1)**
04.3 **This is a model answer, which would gain all 6 marks:**
If the blood is too concentrated, ADH is released by the pituitary gland. This increases the permeability of the kidney tubules and more water is reabsorbed, resulting in less urine and more concentrated urine being produced. When the water level of the blood returns to normal this is detected (by the hypothalamus) which causes the pituitary to reduce ADH production. If the water level of the blood is too high, ADH production stops and a higher volume of dilute urine is produced.
04.4 35% **(1)**
04.5 Heat energy is transferred from the body to its surroundings **(1)** to evaporate the sweat **(1)**.
04.6 Blood vessels dilate **(1)**.
04.7 Enzymes work best at body temperature **(1)**.

TOTAL MARKS FOR QUESTION 4 = 16

05.1 F = no cystic fibrosis   f = cystic fibrosis

		Parent 2	
		F	**f**
Parent 1	**F**	FF	Ff
	f	fF	ff

(1 mark for key; 1 mark for correct genetic cross results; 1 mark for identifying ff as offspring with cystic fibrosis)
05.2 Embryos found to have cystic fibrosis / unused embryos are destroyed, which some people regard as destroying life **(1)**.
05.3 Homozygous **(1)**
05.4 $\frac{1800}{585}$ = 3.076 = 3.08 : 1 to 2 decimal places **(1)**
05.5 Genes had not been discovered **(1)**.

TOTAL MARKS FOR QUESTION 5 = 7

06.1 C **(1)**
06.2 **Any one from:**
• Coordinates voluntary movements
• Posture
• Balance
• Speech **(1)**.
06.3 **Any one from:**
• Biopsy
• MRI scan
• CT scan **(1)**.

06.4 **Any four from:**
- They are slow growing and don't spread.
- Symptoms are not severe…
- … so the risk to the patient is low compared with the risk of surgery.
- Effects of radiotherapy,…
- … particularly in young children, which is where they frequently occur **(4)**.

06.5 Medulla **(1)**

TOTAL MARKS FOR QUESTION 6 = 8

07.1 23 **(1)**
07.2 Testes **(1)**
07.3 Female, because the egg always donates X **(1)**; therefore offspring will have XX which is female **(1)**.
07.4 Mitosis **(1)**
07.5 Advantage – **any one from:**
- Quick
- Only needs one parent
- Can create many clones
- Time- and energy-efficient since no need to search for mate **(1)**.

Disadvantage – **any one from:**
- No variation in offspring means they are all vulnerable if the environment changes unfavourably
- Does not allow natural selection and evolution **(1)**.

07.6 Plants that grow from runners will all be genetically identical to the parent plant **(1)**. The plants that grow from the seeds will be genetically different from each other and the parent plant **(1)**.
07.7 They target the areas where the mosquito lives **(1)**, so there will be less mosquitos to pass on the parasite **(1)** and less mosquitos for the parasite to complete its life cycle **(1)**.

TOTAL MARKS FOR QUESTION 7 = 12

08.1 Biotic – **any two from:**
- Availability of food
- Number of predators
- Number of pathogens **(1)**.

Abiotic – **any two from:**
- Availability of light / temperature / water pH / dissolved minerals / levels of dissolved oxygen or carbon dioxide **(1)**.

(1 mark for one biotic and one abiotic factor)

08.2 **This is a model answer, which would gain all 4 marks:**
Plant plankton begin to increase in March because there is more light for photosynthesis. In April / May shrimp numbers increase because there is more plankton to feed on. Increasing numbers of shrimps eat the plankton so numbers of plant plankton decrease in May. This is followed by decreasing numbers of shrimps in June because their food source is decreasing. This allows plant plankton numbers to increase again in September.

08.3 The deeper the rock pool, the more variety of organisms found **(1)**.
08.4 Shallow rock pools will have greater variation in temperature, which many organisms will not be able to tolerate / Organisms may be easier to spot by predators in shallow pools, so only those with good camouflage will inhabit them **(1)**.

TOTAL MARKS FOR QUESTION 8 = 8

09.1 Gibberellin **(1)**
09.2 The hormone auxin **(1)** is transported to the stem on the side not facing the light **(1)**. The auxin causes the cells on this side to elongate **(1)**, which causes the plant to bend towards the light **(1)**.
09.3 Weedkiller will destroy the food source of some organisms, for example beetles, so their numbers will decrease **(1)**. The beetles in turn could be food for birds, so with no beetles the bird numbers will decrease **(1)**.
09.4 To encourage ripening / to delay ripening **(1)**

TOTAL MARKS FOR QUESTION 9 = 8

10.1 To grow crops for biofuels / to grow staple foods such as rice **(1)**.
10.2 30% **(1)**
10.3 **Any one from:**
- Introduction of fish quotas
- Control of net size
- Increasing mesh size of nets **(1)**.

10.4 **This is a model answer, which would gain all 6 marks:**
Increased levels of carbon dioxide and other greenhouse gases in the atmosphere trap the heat from the Sun when it is reflected from the Earth's surface. This is causing the atmosphere to get warmer, which is causing the ice caps to melt. This results in loss of habitat for animals that live there. Changing sea temperatures affect organisms at the start of the food chain. Global warming also causes changes to rainfall patterns, which means crops may fail due to excess rain or drought. A warmer atmosphere also means that some pests are not killed off by harsh winters and have a longer breeding season.

TOTAL MARKS FOR QUESTION 10 = 9

SET B: PAPER 1

01.1 Root hair cell **(1)**
Any one from:
- Absorbs water / minerals
- Anchorage **(1)**.

01.2 Cell wall **(1)**

01.3 C **(1)**

01.4 Chloroplasts trap / absorb light / are sites of photosynthesis **(1)**. This cell is underground where there is no light / this cell does not carry out photosynthesis **(1)**.

01.5 **Any four from:**
- Increased detail…
- … because there is a high magnification
- … and high resolution
- Can see structures in 3D
- Can see sub-cellular structures **(4)**.

TOTAL MARKS FOR QUESTION 1 = 10

02.1 Movement of water from an area of high water concentration to an area of low water concentration **(1)** across a semi-permeable membrane **(1)**.

02.2 –3, –2, –2.8 **(1 mark for the correct answers; 1 mark for adding the minus signs)**

02.3 Yes (no mark). The cubes did not start at the same mass **(1)** – a bigger cube will gain / lose more water than a smaller cube **(1)**.

02.4 $\frac{3}{15} \times 100$ **(1)** = 20% **(1)**

02.5 The concentration was between 0.2 and 0.4% **(1)**.

02.6 **This is a model answer, which would gain all 4 marks:**
Diffusion is the movement of molecules / particles / substances from an area of high concentration to an area of low concentration, whereas active transport is from an area of low concentration to an area of high concentration. Diffusion does not require energy / is passive, whereas active transport does require energy. Examples of diffusion include oxygen / carbon dioxide exchange between blood and alveoli in lungs; oxygen / carbon dioxide between blood and cells; glucose diffusing from blood to cells (or any other correct example). Examples of active transport include products of digestion / amino acids / glucose from small intestines into blood; mineral ions from soil into plant roots (or any other correct example). **(Only 1 example of diffusion and active transport needed to gain all 4 marks.)**

TOTAL MARKS FOR QUESTION 2 = 13

03.1 It is an artery and carries deoxygenated blood **(1)**.

03.2 It has to generate enough pressure to send blood around the body **(1)**; it therefore has thick muscular walls **(1)**.

03.3 A pacemaker **(1)**

03.4 $A = \frac{4}{5} = 0.8$ **(1)**; $B = \frac{100}{5} = 20$ **(1)**

03.5 Blood vessel B (no mark) because arteries carry blood under high pressure / resistance is high due to smaller lumen **(1)**.

03.6 A **(1)**

03.7 They help the blood to clot **(1)**.

TOTAL MARKS FOR QUESTION 3 = 9

04.1 A = cytoplasm **(1)**; B = cell wall **(1)**

04.2 Sexually **(1)**

04.3 To sterilise it **(1)**

04.4 **This is a model answer, which would gain all 4 marks:**
The inoculating loop should be passed through the flame to sterilise it. When it has cooled it should be dipped into the broth and then spread onto the agar in the petri dish. The loop should be passed through the flame a second time. The petri dish should be sealed with tape and incubated at a suitable temperature.

04.5 6400 **(1)**

04.6 $1600 \times 100 = 160\ 000 / 1.6 \times 10^5$ **(1)** Yes, because 1.6×10^5 is greater than 1×10^5 **(1)**

04.7 Toxins **(1)**

TOTAL MARKS FOR QUESTION 4 = 12

05.1 Carries / binds to oxygen **(1)**

05.2 Nucleus **(1)**

05.3 Undifferentiated cell / cell capable of giving rise to many cells of the same type **(1)**

05.4 Bone marrow **(1)**

05.5 **Any one from:**
- Transfer of pathogens
- Risk of infection **(1)**.

05.6 **Any four from:**
- The drug must be tested in a laboratory…
- … using cells / live animals.
- It should be tested on a small number of people.
- Larger clinical trials on patients and healthy people.
- Double blind trials, in which some patients are given a placebo **(4)**.

TOTAL MARKS FOR QUESTION 5 = 9

06.1 $C_6H_{12}O_6$ **(1)**; $6H_2O$ **(1)**

06.2 Muscles require more oxygen **(1)** to release energy from glucose **(1)**. Breathing faster takes in more oxygen, which is delivered to muscles by blood **(1)**.

06.3 Depth of breathing increases **(1)**.

06.4 It is taken to the liver **(1)** by the bloodstream **(1)** where it is converted to glucose **(1)**.

06.5 Carbon dioxide (no marks) because a small rise in % carbon dioxide gives a large rise in breathing rate / a 7% increase in carbon dioxide increases breathing by 35 l/min **(1)**. A large decrease in oxygen gives only a small rise in breathing rate / a 12% decrease in oxygen increases breathing by about 4 l/min **(1)**. (Other figures from the graphs accepted if feasible.)

06.6 They increase the surface area for diffusion **(1)**.

TOTAL MARKS FOR QUESTION 6 = 12

07.1 Cells produce mucus, which traps pathogens **(1)**. Cilia move trapped pathogens upwards, away from the lungs **(1)**.

07.2 The process is called phagocytosis **(1)**. White blood cells engulf / ingest the pathogen / bacteria **(1)**. They destroy the pathogen using enzymes **(1)**.

07.3 Antibiotics destroy / kill the pathogen causing the disease **(1)**. Painkillers treat the symptoms, e.g. lower the temperature **(1)**.

TOTAL MARKS FOR QUESTION 7 = 7

08.1 Carbon dioxide + water **(1)**; glucose + oxygen **(1)**

08.2 It is being limited by the amount of carbon dioxide available / all the carbon dioxide has been used up **(1)**.

08.3 6 **(1)**

08.4 **Any two from:**
- Respiration
- To produce fat / oil
- To produce cellulose
- To produce amino acids **(2)**.

08.5 **This is a model answer, which would gain all 6 marks:**
The upper epidermis is thin and transparent so light can easily pass through. The palisade cells are packed with chloroplasts for photosynthesis and are near the upper surface of the leaf to catch the light. The spongy mesophyll layer has many air spaces to allow rapid diffusion of gases. The leaf contains veins to supply the plant with water for photosynthesis.

TOTAL MARKS FOR QUESTION 8 = 12

09.1 Health is not only physical. Mental and emotional well-being are also part of being healthy **(1)**.

09.2 Benign – the tumour remains localised / does not spread **(1)**; malignant – the tumour spreads and forms secondary tumours **(1)**.

09.3 Cervical cancer **(1)**

09.4 The higher the weekly meat consumption, the higher the incidence of bowel cancer **(1)**

09.5 F **(1)**

09.6 Allow answers between 1.1 kg and 1.15 kg **(1)**

TOTAL MARKS FOR QUESTION 9 = 7

10.1 $38 - 18 = 20$

$\dfrac{20}{5} = 4$ cm³ **(1)**

10.2 **Any two from:**
- Size / mass of plants
- Length of roots
- Number of leaves
- Size of leaves
- Conditions leaves were left in, e.g. temperature **(2)**.

10.3 **This is a model answer, which would gain all 6 marks:**
In light conditions, the guard cells photosynthesise and produce glucose. This increases the concentration inside the guard cells so they take in water from surrounding cells by osmosis. The cells become turgid. The thin outer wall bends more than the thick inner wall so the stomata open. When photosynthesis stops, the guard cells lose water and the inner walls move close together, closing the stomata.

TOTAL MARKS FOR QUESTION 10 = 9

SET B: PAPER 2

01.1 Third **(1)**

01.2 **Either:**
- Water insects increase because there is less competition for food / more food available.
- Number of frogs increase because there is more food **(2)**.

Or:
- Water insects decrease because they are eaten by the large fish that won't be able to feed on sticklebacks.
- Frog numbers decrease because there are less water insects to eat **(2)**.

01.3 $\dfrac{6 \times 9}{3}$ **(1)** = 18 hedgehogs **(1)**

01.4 No boundaries **(1)** so organisms move freely into and out of the area **(1)**.

TOTAL MARKS FOR QUESTION 1 = 7

02.1 It thickens the lining of the uterus wall **(1)**.

02.2 It releases an egg **(1)**.

02.3 It maintains the lining of the womb **(1)**.

02.4 The pituitary gland **(1)**

02.5 **This is a model answer, which would gain all 6 marks:**
Oral contraceptives are effective if taken as prescribed and are usually free unlike barrier methods such as condoms. However, they do not protect against sexually transmitted diseases whereas some barrier methods do. Oral contraceptives must be prescribed, which can be seen as a disadvantage compared to buying barrier products over the counter. They must also be taken regularly and can have side-effects, unlike barrier methods, which are used only when needed and are unlikely to cause side-effects.

02.6 One or two fertilised eggs will be placed into Fatima's womb **(1)** when they are a tiny ball of cells **(1)**.

TOTAL MARKS FOR QUESTION 2 = 12

03.1 Microorganisms feed on dead and decaying matter **(1)**. The carbon compounds in the matter are used in respiration **(1)**, which returns carbon to the atmosphere as carbon dioxide **(1)**.

03.2 It is converted to glucose by photosynthesis **(1)**, which can then be stored as starch / respired / used to make other carbon compounds **(1)**.

TOTAL MARKS FOR QUESTION 3 = 5

04.1 $\dfrac{(6.24 + 6.21 + 6.16)}{3}$ **(1)** = 6.2 **(1)**

04.2

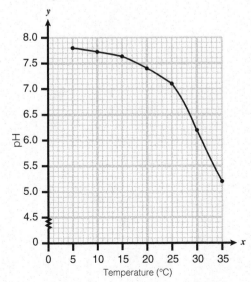

(2 marks for all points plotted correctly or 1 mark for only five points correct; 1 mark for drawing line/curve)

04.3 Between 5°C and 20°C, a rise in temperature causes only a slight increase in the rate of decay **(1)**. Increases in temperature from 20°C to 35°C cause a large increase in the rate of decay **(1)**.

04.4 11–15 days **(1)**

04.5 $\dfrac{20}{80} \times 100 = 25\%$ **(1)**

04.6 Either pig because a higher volume of biogas was produced after 10 days, or cow because the total amount of biogas produced was more after 20 days **(1)**.

TOTAL MARKS FOR QUESTION 4 = 10

05.1 **Any two from:**
- Touch
- Pressure
- Pain
- Temperature **(2)**.

05.2 Glands produce hormones **(1)** and muscles contract **(1)**.

05.3 ⬅ Arrow from right to left **(1)**

05.4 **This is a model answer, which would gain all 5 marks:**
Electrical impulses arrive at the end of the first neurone and cause the release of a neurotransmitter into the synaptic cleft. The neurotransmitter crosses the gap and binds to receptors on the second neurone, which generates a new electrical impulse. If the first neurone is a sensory neurone the impulse will pass to a relay neurone. If the first neurone is a relay neurone the impulse will pass to a motor neurone.

05.5 They protect the body from damage **(1)**.

TOTAL MARKS FOR QUESTION 5 = 11

06.1 A phosphate **(1)**

06.2 TAACGACGGGTA **(1)**

06.3 Four **(1)**

06.4 mRNA copies one strand of the DNA **(1)** and takes it to the ribosomes **(1)** where the correct amino acids are added to the chain **(1)**.

06.5 The gene is switched off **(1)**.

06.6 The wrong protein could be made / the protein might not function correctly **(1)**.

06.7 **Any one from:**
- To find genetic links to disease
- To be able to treat inherited disorders
- To trace human migration patterns from the past **(1)**.

TOTAL MARKS FOR QUESTION 6 = 9

07.1 The gene you want is isolated **(1)** and inserted into a vector / plasmid **(1)**, which is then used to insert the gene into an embryo **(1)**.

07.2 The manure will contain less phytate so will not damage the environment **(1)**.

07.3 Concerns about the safety of consuming genetically engineered food **(1)**.

07.4 They will save money because they won't have to buy phytase to add to pig food **(1)**.

07.5 **This is a model answer, which would gain all 5 marks:**
An egg cell could be taken from a donor pig and the nucleus removed. A body cell would then be taken from the 'enviropig' and the nucleus removed. The nucleus of the body cell would be placed in the empty egg cell and given an electric shock to start cell division. When the egg had become a ball of cells, it would be placed into a pig womb to develop.

TOTAL MARKS FOR QUESTION 7 = 11

08.1 The remains / hard parts / preserved traces / imprints **(1)** of an organism that lived millions of years ago **(1)**.

08.2 Environmental – **any one from:**
- Volcanic eruption
- Asteroid
- Meteorite collision
- Change in climate
- Ice age **(1)**.

Biological – **any one from:**
- New disease
- New predator
- Lack of food
- Habitat destruction **(1)**.

08.3 Variation amongst swordfish meant some had longer swords / mutation in swordfish meant one had longer sword **(1)**. Those with longer swords were able to survive **(1)** and reproduce **(1)**, and pass on genes for longer sword to offspring **(1)**.

08.4 **Any two from:**
- People believed that God created living things.
- Genes had not been discovered.
- There was insufficient evidence **(2)**.

TOTAL MARKS FOR QUESTION 8 = 10

09.1 **Any two from:**
- Increasing birth rate
- New plant pests and pathogens
- Cost of seeds or farming machinery
- War and conflict
- Famine due to crop failure **(1)**.

09.2

Type of farming	Advantages
Intensive (Figure 8)	Cows grow faster because energy is not wasted on movement **(1)** Takes up less space **(1)**
Free range (Figure 9)	Meat tastes better / can sell meat for higher price **(1)** Less chance of disease spreading **(1)**

09.3 All of the crop will flower at the same time, so insects that feed on pollen / flowers will only have food available for a short time **(1)**. Organisms that feed on the insects will not have as much food available **(1)**.

09.4 Plant hedgerows / wild flower borders at field edges **(1)**

09.5 **This is a model answer, which would gain all 4 marks:**
Breed a female taurine cow with a male zebu cow. From the offspring, choose a strong male and a female with good milk production and breed them. Repeat this through many generations.

09.6 Advantage – very fast growing **(1)**; disadvantage – costly to set up / requires technology **(1)**.

TOTAL MARKS FOR QUESTION 9 = 14

10.1 B **(1)**
10.2 The pupil becomes larger **(1)**.
10.3 A **(1)**
10.4 D **(1)**
10.5 When focusing on near objects the lens is fatter and more curved **(1)**. When focusing on far objects the lens is thin and less curved **(1)**.

TOTAL MARKS FOR QUESTION 10 = 6

11.1

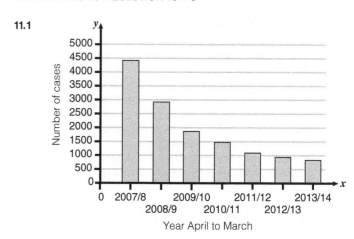

Bar must be drawn to correct height **(1)**

11.2 80–85% **(1)**.
11.3 A mutation causes a bacterium to become resistant **(1)**. Sensitive bacteria are killed by the antibiotic **(1)**. The resistant bacteria reproduce quickly until all the population is resistant **(1)**.

TOTAL MARKS FOR QUESTION 11 = 5

Notes

Notes